DESIGN AND ASSESSMENT OF SUSTAINABLE PRODUCTS

This book questions the current definition of what makes a product sustainable and argues that a holistic approach to sustainable product design is required, one that considers all aspects of a product's life cycle from design to production, to use and then final disposal.

This edited collection introduces a new set of methods and tools aligned with the concept of comprehensive sustainable produce design that integrates the environmental and social benefits of a product in line with the principles of a circular economy. It provides a comprehensive understanding of the theoretical and practical framework that underpins a sustainable product, highlighting the multiple key roles of (eco-)design, innovation, quality, and sustainability. The authors describe the criteria for which products can be defined as being sustainable, and outline how different manufacturing technologies influence the value of those products and the place they can find on the market accordingly. The book's significant contribution lies in identifying the critical factors that are needed to successfully implement the framework throughout the entire life cycle of the product in a holistic integrated approach.

This book will be of interest for researchers and students studying sustainable product design, environmental studies, engineering, and sustainable business management. It will also be a useful resource for representatives of the business community, managers, technicians, decision-makers, and consumers interested in sustainable product design issues.

Magdalena Wojnarowska is an Associate Professor and Head of the Department of Technology and Ecology of Products at the Krakow University of Economics, Poland. She manages and contracts research projects funded by the Minister of Science and Higher Education, the National Science Centre, the National Centre for Research and Development, and the European H2020. Magdalena is an expert in the selection of projects for co-financing from EU funds in the field of environmental protection. She is a representative of the University of Economics in Krakow at KT 270 for Environmental Management of the Polish Committee for Standardization and also in the LifeScience Cluster. She is the author of numerous publications in the field of circular economy, sustainable production, sustainable consumption, life cycle assessment, and eco-design. Her research focuses on circular economy, bioeconomy, environmental management tools, sustainable production, and sustainable consumption.

Carlo Ingrao is a tenure-track Assistant Professor with associate professorship habilitation in 'Commodity Science' in the Department of Economics, Management, and Business Law of the University of Bari Aldo Moro, Italy. He obtained an M.Sc. degree in Engineering for the Environment and the Territory at the University of Catania and has twice been awarded the title of PhD: one in 'Geotechnical Engineering' at the University of Catania and another in 'Civil Infrastructures for the Territory' at the Kore University of Enna. Ranked among the top 2% of the world's most cited and influential researchers for 2021, his main research activities and interests focus on sustainability, circular economy, and life cycle assessment applications in a wide range of sectors. Within those themes, he has authored around 90 publications in indexed international journals, book chapters, and conference proceedings, has taught university courses, and has delivered numerous lectures and seminars.

DESIGN AND ASSESSMENT OF SUSTAINABLE PRODUCTS

A Conceptual and Practical Framework

Edited by
Magdalena Wojnarowska and Carlo Ingrao

Routledge
Taylor & Francis Group

LONDON AND NEW YORK

from Routledge

First published 2025
by Routledge
4 Park Square, Milton Park, Abingdon, Oxon OX14 4RN

and by Routledge
605 Third Avenue, New York, NY 10158

Routledge is an imprint of the Taylor & Francis Group, an informa business

British Library Cataloguing-in-Publication Data
A catalogue record for this book is available from the British Library

ISBN: 978-1-032-71068-6 (hbk)
ISBN: 978-1-032-71067-9 (pbk)
ISBN: 978-1-032-71069-3 (ebk)

DOI: 10.4324/9781032710693

Typeset in Times New Roman
by codeMantra

CONTENTS

FIGURES

TABLES

CONTRIBUTORS

Ignazio Blanco earned his degree in Industrial Chemistry and his PhD in Polymeric Materials for special uses at the University of Catania. Since December 2018, he has been a Full Professor of Chemical Foundations of Technologies in the Department of Civil Engineering and Architecture of the University of Catania, where he also serves as the coordinator of the degree course in Civil, Environmental, and Management Engineering. Since 2019, he has held the position of Italian Affiliate Councilor at the International Confederation for Thermal Analysis and Calorimetry.

Pierluigi Catalfo is a Professor of Business Economics and Accounting at the University of Catania. He is the Director of the Research Center for Governance of Territorial Development (GOT) at the University of Catania and the President of the degree course in Management for Sustainable Economy. He teaches financial and management accounting and strategic planning and control for public sector. His main research focus is on intangible values.

Massimo Riccardo Costanzo is a Research Fellow in 'Commodity Sciences' in the PNRR project 'On Foods'. He holds a degree in Economics from the University of Catania and a PhD from the University of Messina, where he also obtained a degree in Tourism and Environmental Management. He is involved in research in the fields of voluntary certifications, multi-criteria aggregation in sustainability, and corporate social responsibility.

Artur Jachimowski is an Assistant Professor in the Department of Technology and Ecology of Products at the Krakow University of Economics. He is also a manager and contractor for national research projects. Publications are in the fields of environmental engineering, municipal economy, municipal waste management, waste logistics, and multi-criteria analysis methods.

Szymon Jarosz is a Research and Teaching Assistant in the Department of Technology and Ecology of Products at the Krakow University of Economics. He participates in research projects and contributes as an author of scientific papers. His research interests focus on innovation, sustainable development, sustainable transformation of enterprises, and environmental management. He is the recipient of Laureate of the Best Student of the Institute of Management at the Krakow University of Economics Award in 2020 and the prestigious Students' Nobel Prize Award in Economics in 2021.

Agata Matarazzo is an Associate Professor of Commodity Sciences in the Economic & Business Department at the University of Catania (Sicily), where she teaches Technology of Sustainability Productions; and Quality, Environmental, and Safety Management Systems. Her research focuses on quality and environmental management systems, life cycle assessment, circular economy, eco-sustainability instruments, air pollution impacts, new technologies in energy production plants, and corporate social responsibility.

Alina Matuszak-Flejszman is a Professor of Social Science, Director of the Institute of Management, and Head of the Department of Quality Management at the Poznań University of Economics. She manages and contracts research projects financed by the Ministry of Science and Education and serves as an expert in the assessment of research projects of the National Centre for Research and Development. Her publications focus on environmental management, quality management, risk management, and sustainable production. In addition, she is a consultant, advisor, and trainer.

Karolina E. Mazur is a Research Assistant in the Department of Technology and Ecology of Products, Krakow University of Economics and at the Faculty of Materials Engineering and Physics Krakow University of Technology. Her main research interests include the production of functional polymer composites with enhanced mechanical properties and with a low negative impact on the natural environment.

Magdalena Muradin (PhD, Economics) is an Assistant Professor in the Department of Quality Management at the Poznań University of Economics and Business, Poland. Her academic and research areas include sustainability, environmental management, life cycle assessment (LCA), circular economy management, and project management. She is also involved as an LCA and circular economy expert in European, national, and regional projects in the field of biomass, bioenergy, and construction.

Tomasz Nitkiewicz (PhD, Management) is an Associate Professor in the Department of Management and Entrepreneurship, at the Częstochowa University of Technology, Poland. He is a founder and the current Head of the Life Cycle Modeling Centre at the Częstochowa University of Technology. He is also engaged in initiatives in developing education in management and engineering towards meeting the new challenges of Industry 4.0 and the circular economy (MSIE4.0 project). He is a member of Natureef Association, which bring together packaging producers, food producers, recyclers, researchers, and other stakeholders of packaging value chain, where he acts as ESG, LCA, and carbon footprint expert.

Beata Paliwoda is a researcher, consultant, and auditor in Environmental and Quality Management. She is a Lecturer at the Poznan University of Economics and Business and manages and contracts various research projects. She has also managed complex projects in automotive, railway, and aviation industries and is an auditor for ISO 9001, ISO 14001, and AS 9100 standards. Her published works focus on the effectiveness and performance of quality management systems, environmental management systems, Industry 4.0, and Industrial Internet of Things.

Marcin Paprocki is a mechanical engineer and an Assistant Professor in the Department of Technology and Ecology of Products at the Krakow University of Economics. He is also a contractor of national research projects. His publications are in the fields of management and production engineering, business and engineering process modeling, designing and improving product and production processes, Six Sigma, Lean Sigma, sustainable design, and sustainable product development.

Marcin Rychwalski is an Assistant Professor in the Department of Technology and Ecology of Products at the Krakow University of Economics. He is a participant and contractor of numerous scientific projects regarding optimizing the environmental footprint of products and processes using the life cycle assessment (LCA) method. His publications and research interests include environmental impacts and footprints using LCA, sustainable products, and the circular economy.

Mariusz Sołtysik is a Professor at the University of Economics in Krakow and the Coordinator of implementation works in the Strategic Scientific Research and Development Program. He manages and contracts projects financed by the National Science Centre and the National Centre for Research and Development, including the Leader program, and is also a contractor in the H2020-EU program. His main research interests include strategy, innovation, dynamic capabilities, and circular strategies.

Martin Straka is a Professor at the Institute of Logistics and Transport at the Technical University of Kosice, where he also serves as the Director. His research interests are in the areas of distribution and supply logistics, computer simulation, production logistics, and information logistics.

Sergiusz Strykowski is in the Department of Information Technology at the Poznań University of Economics. He specializes in IT projects related to machine learning and intelligent data processing. He has served as a technological and R&D leader, an IT systems architect, and a lead software developer in over 30 R&D, scientific, and industrial projects.

Erica Varese (PhD) is currently an Associate Professor in Commodity Science in the Department of Management, University of Torino, Italy, where she teaches undergraduate, graduate and master's courses in Italian and English. Her publications are in the fields of circular economy, sustainable production and sustainable consumption, consumer protection and perception, and international trade and customs.

Tomasz Witko is an experienced researcher and expert in the field of biotechnology and biopolymers. He is an Assistant Professor in the Department of Technology and Ecology of Products at the University of Economics in Krakow, specializing in ecology and sustainable product development and new additive manufacturing techniques. His extensive scientific achievements include participation in significant grants from NCN and NCBiR, as well as publications in reputable journals. He actively collaborates with the industrial sector.

Mateusz Wygoda is a mechanical engineer and Assistant Professor in the Department of Technology and Ecology of Products at the Krakow University of Economics. He has participated in scientific projects related to technical issues and rationalization of production processes. His publications are in the fields of solid mechanics, stress concentration factor, fatigue life, destructive and non-destructive testing methods, strain measurement, and technical issues in the manufacturing process.

ABBREVIATIONS

10R	(Refuse, Rethink, Reduce, Reuse, Repair, Refurbish, Remanufacture, Repurpose, Recycle, Recover)
3R	(Reduce, Reuse, Recycle)
ADF	Abiotic Depletion Factor
AHP	Analytic Hierarchy Process
AI	Artificial Intelligence
AM	Additive Manufacturing
ANP	Analytic Network Process
APS	Atmospheric Plasma Spray
BDA	Big Data Analytics
BOCR	Benefits, Opportunities, Costs, Risks
BPMN	Business Process Modeling Notation
CAD	Computer-Aided Design
CAE	Computer Aided Engineering
CAM	Computer-Aided Manufacturing
CCB	Cepi Container Board
CE	Circular Economy
CEM	Cement Production Classification
CERES	Coalition for Environmentally Responsible Economies
CF	Carbon Footprint
CMU	Circular Material Use
CR	Consistency Ratio
CS	Cold spray
CSR	Corporate Social Responsibility
CSRD	Corporate Sustainability Reporting Directive
DFA	Design for Assembly

DFD	Design for Disassembly
DFE	Design for Environment
DFL	Design for Longevity
DFM	Design for Manufacture
DFMA	Design for Manufacture and Assembly
DFR	Design for Recycling
DFRem	Design for Remanufacturing
DFromR	Design from Recycling
DFS	Design for Sustainability
DFX	Design for X
DNA	Deoxyribonucleic acid
DOE	Design of Experiments
ECD	Environmentally Conscious Design
EFRAG	European Financial Reporting Advisory Group
E-LCC	Environmental Life Cycle Costing
EMAS	Eco-Management and Audit Scheme
EMS	Environmental Management Systems
EN	European Standard
EoL	End-of-Life
EPD	Environmental Product Declaration
EPR	Extended Producer Responsibility
EQFD	Environmental Quality Function Deployment
ESG	Environment, Social Responsibility, Corporate Governance
ESRS	European Sustainability Reporting Standards
EU	European Union
EU ETS	European Union Emissions Trading System
EUR	Euro
FEFCO	Fédération Européenne des Fabricants de Carton Ondule
FMEA	Failure Mode and Effects Analysis
GRI	Global Reporting Initiative
GWP	Global Warming Potential Factors
HVOF	High speed oxyfuel
IIoT	Industrial Internet of Things
IIRC	International Integrated Reporting Council
IoT	Internet of Things
IPCC	Intergovernmental Panel on Climate Change
IS	Industrial Symbiosis
ISSB	International Sustainability Standards Board
JRC	Joint Research Centre
LCA	Life Cycle Assessment
LCC	Life Cycle Cost
LCI	Life Cycle Inventory

LCIA	Life Cycle Impact Assessment
LCSP	Lowell Center for Sustainable Production
M2M	Machine-to-Machine
MCDM	Multi-Criteria Decision-Making
MCI	Material Circularity Indicator
NFCP	National Energy and Climate Plan
NFRD	Non-Financial Reporting Directive
NPD	Developing New Products
OEM	Original Equipment Manufacturing
PBAT	Poly(Butylene Adipate-co-Terephthalate)
PCL	Polycaprolactone
PDCA	Deming Cycle (Plan, Do, Check, Act)
PE	Polyethylene
PEF	Product Environmental Footprint
PET	Polyethylene Terephthalate
PGGA	Poly(Glycolic Acid-co-Glycolic Acid)
PHAs	Polyhydroxyalkanoates
PHB	Polyhydroxybutyrate
PIEs	Public Interest Entities
PLA	Polylactic Acid
PLGA	Poly(Lactic-co-Glycolic Acid)
PMMA	Polymethyl Methacrylate
PN	Polish Standard
PP	Polypropylene
PTT	Poly(Trimethylene Terephthalate)
PVC	Polyvinyl Chloride
QFD	Quality Function Deployment
QMI	Quality Management Innovation
QMS	Quality Management System
RE	Reverse Engineering
RFID	Radio Frequency Identification
RM	Rapid Manufacturing
RP	Rapid Prototyping
RP/RT	Rapid Prototyping/Rapid Tooling
rPET	recycled Polyethylene Terephthalate
RPN	Risk Priority Number
RT	Rapid Tooling
SASB	Sustainability Accounting Standards Board
SBTs	Science Based Targets
SC	Supply Chain
SCP	Sustainable Consumption and Production
SDG	Sustainable Development Goal

SDGs	Sustainable Development Goals
SFRD	Sustainable Finance Disclosures Regulation
S-LCA	Social Life Cycle Assessment
SME	Small and Midsize Enterprise
SO	International Organization for Standardization
SOPs	Standard Operating Procedures
SP	Sustainable Procurement
SPIs	Sustainable Production Indicators
SSC	Sustainable Supply Chain
SSCM	Sustainable Supply Chain Management
TBL	Triple Bottom Line
TCFD	Task Force on Climate related Financial Disclosures
UN	United Nations
UNEP	United Nations Environment Programme
US	United Sates
VoC	Voice of the Customer
VOCs	Volatile Organic Compounds
VRE	Value-based Resource Efficiency
WH	Waste Hierarchy
WM	Waste Management

INTRODUCTION

Design and assessment of sustainable products: a conceptual and practical framework

Magdalena Wojnarowska and Carlo Ingrao

Traditional products are evolving towards sustainability, making it increasingly crucial to consider environmental and social factors in the design process. Sustainable product design can address this need by integrating environmental and social aspects at all stages of the product life cycle, alongside more conventional business, technical, and financial aspects. This holistic approach ensures that products are not only economically viable and technically sound but also environmentally friendly and ethically responsible. By doing so, sustainable product design aims at minimising negative impacts on the environment and society while maximising positive outcomes.

As society becomes more environmentally conscious and knowledge-based, the circular economy is gaining importance. Consumers are increasingly seeking information about the quality, safety, and sustainability of the products they purchase. They want assurance that the products they buy do not harm the environment and contribute positively to society. This shift in consumer behaviour drives the need for transparency and accountability in the entire product lifecycle, from the raw material extraction to end-of-life disposal, passing through the phases of commodity making, distribution, and use. Therefore, according to the editors and the authors of this book, there is a need for methods that enable the preliminary consideration and assessment of such important issues at the conceptual and design stages. This approach allows designers and manufacturers to understand and mitigate potential negative impacts in the earliest stages of design and development, ensuring that sustainability is built into the product from the beginning.

Designing sustainable products requires a wide range of tools and methods to study the product and its features throughout its entire life cycle, both from a theoretical and a practical perspective. This necessitates a holistic approach, considering all aspects of the product's life cycle from design to production, use, and final disposal. It involves the application of the life cycle assessment (LCA) approach

DOI: 10.4324/9781032710693-1

to evaluate not only the environmental burdens of the product's life cycle but also the economic and the social ones, making sure that the product itself is beneficial for all stakeholders involved. Furthermore, it requires the integration of innovative technologies and materials that enhance sustainability, such as biodegradable and recycled materials and energy-efficient processing technologies.

This book reviews and builds upon that approach in a holistic integrated approach, to fill the research gap that both the editors and the authors have found in defining sustainable products and providing comprehensive tools and research methods for their design and assessment. These tools and methods allow for a reliable description of the environmental and social features of a sustainable product in line with the principles of a circular economy that meets the goals of sustainable development. By incorporating these principles, this book provides a framework for creating products that contribute to a more sustainable and equitable world.

In the context of the global ecological transition, this book responds to the need to fill the current research gap related to the insufficiently defined theoretical and practical framework of what a sustainable product is and what methods and tools can be used for its design and assessment throughout its entire life cycle. It provides, in fact, a comprehensive understanding of that framework, highlighting the multiple key roles of eco-design, innovation, quality, and sustainability. A significant contribution of the book stays in identifying the critical factors needed to successfully implement the aforementioned framework in a holistic and integrated manner. By doing so, it offers a roadmap for companies and designers to follow in their journey towards sustainability.

Throughout the 14 chapters, this book attempts to describe the criteria defining sustainable products, and to explain how different manufacturing technologies influence the value of those products and their place on the market. This book reviews and builds upon existing approaches, strategies, methods, tools, legal regulations, and other related issues. In doing so, it contributes to facilitating the modelling, comparison, and balance of the technical, business, financial, environmental, and social aspects of products according to the principles of the circular economy, sustainable development, and corporate social responsibility (CSR). Such a holistic approach ensures that all aspects of sustainability are considered, leading to the development of products that are not only environmentally friendly but also economically viable and socially beneficial.

In the context of increasing plastic consumption across all life sectors and their well-documented negative environmental impacts, this book emphasises the need for innovation and improvement. Under this perspective, focusing on the development and market introduction of innovative biopolymers and biopolymer-based products is an effective strategy.

The content of this book comprises 14 chapters prepared by 21 authors specialising in areas related to sustainable products, including design, manufacturing, marketing, business models, and consumption.

The first chapter delves into the sustainable transformation of enterprises, discussing strategies and practices for achieving sustainability in business operations. Chapter 2 focuses on the transformation of conventional products into sustainable ones, offering guidelines and methodologies for this critical shift. In Chapter 3, the processes involved in designing and developing sustainable products are explored, highlighting key principles and innovative approaches. Chapter 4 examines the role of LCA in the design and development of sustainable products, demonstrating how LCA can be used to improve environmental performance. Chapter 5 covers sustainable production, discussing methods and practices for manufacturing products in an environmentally friendly manner. Chapter 6 addresses the concept of a sustainable supply chain, exploring strategies for making supply chains more sustainable from sourcing to delivery. Chapter 7 investigates the impact of digitalisation on sustainable products and production, showcasing how digital technologies can enhance sustainability efforts. Chapter 8 focuses on waste management in sustainable development, providing insights into effective waste reduction and recycling strategies. Chapter 9 discusses the role of CSR in the sustainability of products, highlighting how CSR initiatives can drive sustainable practices. Chapter 10 looks at marketing sustainable products, exploring how to effectively promote and sell products that are environmentally friendly. Chapter 11 examines sustainable consumption of products, analysing consumer behaviour and strategies to encourage more sustainable consumption patterns. Chapter 12 presents a case study on sustainable product innovation, specifically focusing on biopolymers and their potential as sustainable materials. Chapter 13 further explores the LCA of biopolymers, providing a detailed analysis of their environmental impact and benefits as sustainable products. Chapter 14 conducts a cost–benefit analysis of the biopolymer lifecycle, evaluating the economic and environmental trade-offs involved in using biopolymers as sustainable alternatives.

The authors believe that, in its current form, this book will attract a diversified audience, including representatives of the business community, managers, technicians, decision-makers, and consumers interested in sustainable product design issues, as well as researchers, students, and practitioners in the field of sustainable product development.

Finally, by addressing the theoretical and practical aspects of sustainable product design, this book offers valuable guidance and inspiration for those committed to making a positive impact on the environment and the society.

Acknowledgement

The publication/article presents the result of the Project no 070/ZJE/2024/POT financed from the subsidy granted to the Krakow University of Economics.

1

THE SUSTAINABLE TRANSFORMATION OF ENTERPRISES

Magdalena Wojnarowska, Mariusz Sołtysik,
Tomasz Nitkiewicz and Szymon Jarosz

1.1   Introduction

Although a shift in thinking about the environment and its relationship with socio-economic development first took shape back in the 1960s, the linear economy model – focused primarily on the extraction of cheap, readily available raw materials and energy, with no plans for waste reuse – remains dominant. According to Özkan and Karataş Yücel (2020), this model leads to losses in the material value chain, shorter product life cycles, overproduction of waste, and consequently, ecosystem degradation. Due to these factors, the linear economy model is considered to be both environmentally and socioeconomically unsustainable (Korhonen, Honkasalo, & Seppälä, 2018). Over time, several problems with this economic model have been documented, including:

- the extraction of raw materials exceeds the rate of their natural regeneration,
- following their use, products often end up in landfills or are processed in incinerators, which leads to wastage of valuable and scarce natural resources that must be re-extracted to produce new items.
- hazardous waste management, typical of the linear economy model, which generates pollutants that leach into the soil, water, and air, thus creating alarming levels of environmental pollution.
- both the production and transport of products contribute significantly to energy consumption and environmental pollution.

Therefore, states and markets are striving to respond effectively to the pressing challenges of the present day, such as climate change, social injustice, discrimination, and poverty (Kickul & Lyons, 2020). Throughout human history, numerous

DOI: 10.4324/9781032710693-2

attempts have been made to ensure social and economic security for all members of society. Nevertheless, despite the existence of many models, inequalities and insecurities persist in both society and the economy (Chancel et al., 2022). As early as half a century ago, at the Stockholm Conference on the Environment in 1972, concerns were raised over the need for a shared vision and principles that could inspire and guide the peoples of the world to protect and improve the human environment (Štrukelj, Dankova, & Hrast, 2023).

In view of the above, it is evident that reducing the negative impact of corporate activities on the environment has become a key challenge not only for public administrations but also for private entrepreneurs (Laner & Rechberger, 2009; van Berkel & Kortman, 1993). Overcoming this challenge requires transforming the current economic model into a different, alternative way of doing things, one that seeks to harmonize economic growth with environmental protection needs and various social factors.

1.2   Concepts of sustainable development and the circular economy in the activities of enterprises

Alternative economic models, such as the concepts of sustainable development and the circular economy (CE), have gained in importance in recent times and are now among the most discussed and studied approaches to achieving the Sustainable Development Goals (SDGs; Merli, Preziosi, & Acampora, 2018). A link exists between these concepts, as confirmed by several studies, such as Ruiz-Real et al. (2018), who identify eco-innovation, eco-design, waste management, and the concept of sustainability as the main trends in CE research. In their research, Cecchin et al. (2021) point out that the concept of the CE emphasizes a system of production and consumption that keeps the input and output cycles within a closed loop, making it regenerative and restorative. This approach could contribute to the goal of achieving a "sustainable future". Nitkiewicz (2021) argues that circularity stresses the importance of economic and environmental factors, while social factors play a role indirectly and with less significance and complexity compared to the concept of sustainability. Nevertheless, despite numerous studies analysing the relationship between the CE and sustainable development, it is still unclear how the CE can promote economic growth while taking care of the environment and ensuring social equality within and between generations, as noted by Millar, McLaughlin and Börger (2019).

The increasing pace of environmental and technological change is exerting a considerable influence on the growth patterns of companies looking for new ways to achieve their strategic goals while ensuring long-term success (Wojnarowska, Sołtysik, & Ingrao, 2022, Sołtysik et al., 2024). One of the principal ways to harmonizing economic, social, and environmental goals in enterprises is to implement a business model based on the assumptions of sustainable development and the CE, which represents one of today's greatest challenges (Lüdeke-Freund, Gold, &

Bocken, 2019). For companies, sustainability means implementing strategies and business activities that meet the current needs of the company and its stakeholders, while protecting, maintaining, and improving the human and natural resources that will be needed in the future (Ketprapakorn & Kantabutra, 2019). The main objectives of implementing a CE strategy in an organization include reducing the amount of primary materials and waste production (Haas et al., 2015; Smol & Kulczycka, 2019), as well as promoting environmental protection and pollution prevention (Ma et al., 2014). The common goal of sustainable development and the CE is to enable economic growth while optimizing the use of natural resources through a profound transformation of production chains, consumption patterns, and the redesign of industrial systems. For these concepts to generate synergistic potential, they need to be integrated into business planning, measurement, and management systems (Caiado et al., 2017).

Recent years have witnessed a recalibration of the approach to corporate governance and sustainability research, marked by a shift away from a more conceptual-based perspective towards more strategic and practical analyses (Pandey et al., 2023). Current research shows a positive relationship between organizational management strategies and their sustainable performance and development (Suriyankietkaew & Kungwanpongpun, 2022). Evidence suggests that adopting sustainability principles can bring organizations numerous benefits, such as improved organizational effectiveness and competitiveness, as well as an enhanced reputation (Hristov, Chirico, & Ranalli, 2022). Therefore, in today's global environment, resilient organizations must adopt sustainable management models (Naciti, Cesaroni, & Pulejo, 2021) if they wish to achieve at least one or as many SDGs as possible and increase their chances of success (Hristov, Chirico, & Ranalli, 2022). To achieve this goal, they must acquire the right knowledge and focus on innovation (Štrukelj, Dankova, & Hrast, 2023).

1.3   Transforming the business model into a sustainable enterprise

The phenomenon of "transformation", that is, significant changes in the conditions in which decisions are made and actions are taken, has gained particular importance in recent years in various fields of science, including in the discipline of management (Zgrzywa-Ziemak, 2019). In the face of significant global climate change and environmental degradation, the transition to a sustainable business model is becoming increasingly important, but it comes with enormous challenges faced by economic, political, and social actors (Jansen, 2003). These challenges stem in part from rapid technological advances that have increased organizational efficiency in recent decades, while at the same time raising expectations regarding business operations based on sustainable principles (Ruiz-Mercader, Meroño-Cerdan, & Sabater-Sánchez, 2006). This is why sustainability is increasingly becoming a key factor in achieving competitive advantage from an economic, social, and environmental perspective. It requires enterprises to engage in business activities largely driven by the new strategies. Sustainable business strategies are also key to developing the right capabilities needed to drive sustainable business transformation (Gupta, 2020; Sołtysik et al., 2024).

The main challenge facing organizations today is the need to consider all aspects of sustainable development and to integrate them into their own definitions of growth, in particular their visions, business policies, strategies, structures, and development programmes, as well as take advantage of external opportunities in the currently changing global arena (Heras-Saizarbitoria, Urbieta, & Boiral, 2022). If an organization's activities are based on clear development guidelines and growth is focused on sustainable development, it can be assumed that this organization's activities will also be sustainable (Al-Jayyousi et al., 2023).

Unlike traditional entrepreneurship, the main goal of which is to generate profit, sustainable entrepreneurship does not focus solely on economic success. It also aims to have a positive impact on society and the environment (Halberstadt, Schwab & Sascha Kraus, 2024).

It is worth noting that the one specific area in which a social and environmental impact is generated plays a vital role in the context of sustainable entrepreneurship. Some studies still focus on sustainability solely in ecological terms, assigning the term "green" to sustainable entrepreneurship (Koe, Omar, & Sa'ari, 2015). However, current research is leaning towards a more holistic understanding of sustainability (Sarango-Lalangui, Santos, & Hormiga, 2018). One line of argument suggests that entrepreneurship can only be considered sustainable if entrepreneurial activities (an economic perspective) are geared towards both social and environmental benefits (social and ecological perspectives) (Parrish & Foxon, 2006).

Researchers have discussed different forms of sustainable entrepreneurship, and argue that sustainable entrepreneurship has the potential to generate significant benefits. Sustainable entrepreneurship plays a significant and noteworthy role in promoting SDGs (Sołtysik et al., 2024). Through the creation of new businesses, it is seen as an important factor in ensuring employment, reducing inequality, alleviating poverty, and fostering sustainable growth (Ratten et al., 2019).

As with entrepreneurship itself, entrepreneurial intentions can be closely linked to SDGs in a variety of ways, contributing to initiatives that address socio-economic and environmental issues (Pereira, Rodrigues, & Veiga 2024). Thus, entrepreneurial intentions can help promote the achievement of various goals and objectives set out in SDGs, which thus offers a dynamic approach to tackling social and environmental challenges (Malhotra & Kiran, 2023).

According to Cohen and Winn (2007), sustainable entrepreneurship is based on individuals identifying, developing, and seizing opportunities to create future goods and services that provide economic, social, and environmental benefits. While some researchers still see it primarily as a response to environmental challenges, most definitions emphasize a more comprehensive concept of sustainability. As a consequence, sustainable entrepreneurship is understood as the use of entrepreneurial activity in creating and implementing innovative ideas aimed at solving social and/or environmental problems, while aiming at both achieving maximum social benefits and maximizing profits (Spiegler & Halberstadt, 2018).

As Farny and Binder (2021, p. 605) put it: "Sustainable entrepreneurship describes the nexus between sustainable development and entrepreneurship. Entrepreneurs aspire to create viable market solutions and intend to act as change agents who realize and exploit opportunities for sustainable development." According to Patzelt and Shepherd (2011, p. 137)

> [s]ustainable entrepreneurship is focused on the preservation of nature, life support, and community in the pursuit of perceived opportunities to bring into existence future products, processes, and services for gain, where the gain is broadly construed to include economic and non-economic gains to individuals, the economy, and society.

Sustainable corporate transformation refers to the ability of companies to acquire and rebuild both internal and external organizational resources, including in environmental (ecological) and technological terms. Its goal is to enable enterprises to adapt to the changing business environment by embracing new pathways and creating innovative strategic practices (Wohlgemuth & Wenzel, 2016). These transformation processes actively support circular strategies and eco-innovation practices in a dynamic business environment, providing an important source of competitive advantage for companies (Dangelico, 2016). The main goal of sustainable companies is to strike a balance between economic benefits and environmental responsibility, with the emphasis on pursuing the strategic goal of sustainable development. This means that the transition to a new economic model will require entrepreneurs to adopt a strategic approach that treats environmental objectives as an integral part of their long-term business strategy. Seroka-Stolka (2023) points out that this goal could be achieved in a complex and timely manner solely by means of a proactive approach. It will also require a management philosophy based on a variety of environmental management systems and tools that will enable a company to monitor, evaluate, and improve its environmental performance (Nowicki et al., 2023). This, in turn, will make it possible for companies to effectively integrate economic goals with certain aspects of sustainability, which is key to the task of building a future-proof and sustainable economy (Sołtysik, Urbaniec, & Wojnarowska, 2019; Sołtysik et al., 2024).

Strategic environmental protection measures are playing an increasingly important role in the efforts of companies to gain a competitive advantage. Despite suggestions made by some researchers that strategies supporting sustainability can generate additional costs and reduce the productivity of companies (Palmer, Andres, & Kumar, 1995), other researchers explain that the same strategies can have a positive impact on a company's competitive advantage by using more efficient processes, increasing productivity, reducing costs, and generating new opportunities on the market (Porter & Linde, 1995; Wagner, 2009). Switching to a sustainable business model can therefore be a key success factor for companies, while at the same time contributing to environmental protection and economic efficiency.

Developing a pro-environmental strategy involves many factors, including legal regulations, policy support, green incentives on the market, increasing interaction with consumers by offering green products, creating and diffusing knowledge and technology, entrepreneurial opportunities, and organizational culture (Cohen, 2006; Kiefer, Del Río González, & Carrillo-Hermosilla, 2019; Muñoz & Cohen, 2018; Niemann, Dickel, & Eckardt, 2020; Sołtysik et al., 2019; Volkmann et al., 2021). These factors can be divided into internal and external categories, reflecting the traditional division of an enterprise's environment into micro, meso, and macro levels. This means that the choice of a company's development strategy is based on the interaction between the different socio-economic levels that form an interconnected hierarchy (Johnson & Schaltegger, 2020; Salmivaara & Kibler, 2020). Depending on the stage of a company's development, market position, or technological advancement, it will strive to develop and implement pro-environmental development strategies (Falcone et al., 2020; Kuckertz, 2020). These strategies are based on an assessment of potential business prospects and innovation potential, with the growing importance of innovation placing new demands on pursuing effective strategies that can exploit the innovation potential of enterprises (Sołtysik et al., 2024).

Achieving sustainability greatly depends on a company's ability to ensure comprehensive and integral management, in which owners and managers are aware of sustainability issues. Elements of sustainability must be woven into a company's vision, business policy, core, and other strategies, and their implementation is crucial. When analysing sustainability-related achievements, it is important to consider the individual level as a starting point that shapes various social, economic, and environmental aspects in a company as well as the macroeconomic levels do. This perspective, known as the triple benefit approach (people, planet, and profits), promotes individual accountability, leading to individual competitive advantage, followed by organizational accountability (Lucendo-Monedero, Ruiz-Rodríguez, & González-Relaño, 2023). After all, achieving sustainable development at the global level requires making responsible decisions and actions at all levels – at the individual and the organizational levels, and right through to the macroeconomic level. Governors, managers, and leaders must incorporate social, environmental, and economic benefits into their decisions to guarantee the survival of our civilization in the face of limited natural resources (Liao, 2022).

1.4   Environmental management and sustainability reporting in enterprises

Environmental management has become a key feature of the strategies of many modern enterprises, especially in the context of embracing the ideas of sustainable development (Baumgartner, 2014). In the face of global environmental challenges, companies are looking for ways to be sustainable when it comes to both their business goals and their responsibility to the planet (Ikram et al., 2019).

The concept of environmental management in enterprises is often envisaged as a transparent, systematic process that is recognized across the corporate structure. It is primarily aimed at prescribing and implementing environmental objectives, policies, and responsibilities (Pacana, 2018). Furthermore, this concept also involves regular auditing of its various components to ensure compliance with, and efficacy in, meeting set environmental standards. This approach reflects a holistic commitment to environmental stewardship and sustainable practices within a company's organizational framework (Steger, 2000). In other words, the idea is to incorporate environmental issues into a company's decision-making processes so that they become an integral part of its business (Dey et al., 2018).

Environmental management is no longer just a reaction to external pressures or increasingly stringent laws regulating the environmental impact of a company and its activities. It is also a strategic choice made by companies that understand the value of sustainability and the long-term benefits of environmental responsibility (Wang & Sarkis, 2017). More companies today recognize that environmental responsibility can bring them tangible business benefits, such as reduced costs, innovation, and an improved brand image (Ferrón Vílchez & Darnall, 2016).

One of the best-known standards introduced for environmental management systems is ISO 14001 "Environmental management systems. Requirements with guidance for use" (de Oliveira Neves, Salgado, & Beijo, 2017), which was first issued in September 1996. In the European Union, by ensuring that its system meets the requirements of ISO 14001 certification, a company is regarded as having taken a step towards registration in the Eco-Management and Audit Scheme (EMAS) system, which is one of the most important instruments of the European Union's environmental policy. For several years now, ISO 14001 certificates have been regarded as one of the means by which an organization can manage its impact on the environment and as a consequence help achieve its sustainable development environmental goals (Bravi et al., 2020). Possession of a certificate is no longer seen simply as an additional weapon for winning tenders or as a marketing tool. The absence of an implemented management system based on the ISO standard is often perceived in a negative light and, in many cases, can be seen as barrier to establishing business contacts (Psomas, Fotopoulos & Kafetzopoulos, 2011). Complying with the requirements of the standard is an important step bringing an organization up to a global level of management, in which environmental management plays an important role.

The ISO 14001 environmental management system is an effective tool for ensuring continuous improvement and optimization of a company's organizational structure in the field of environmental protection (Muktiono & Soediantono, 2022). Within this system, several key elements can be distinguished, the first of which is an environmental policy that reflects the company's commitment to using resources efficiently, protecting environmental values, and preventing pollution (Lee et al., 2017). This facet, in turn, is followed by planning, which involves an ecological assessment of the company's operations, an analysis of legal requirements, and the creation of an environmental management programme. The next stage, comprising implementation and functioning, involves creating a well-documented organizational structure with a clear division of positions and responsibilities, organizing training, and preparing for crisis situations and supervising documentation (Pacana, 2018). Another important aspect is control and corrective actions, including monitoring the effectiveness of environmental activities, conducting regular reviews, and assessing compliance with legal regulations (Tudoran, Marin, & Condrea, 2020). The purpose of the management review is to verify and assess the effectiveness of an organization's environmental management system. The ISO 14001 standard of 2015 is based on the PDCA methodology, which stands for plan, do, check, and act (Ahmed & Mathrani, 2021). This methodology focuses on setting objectives and processes in accordance with the organization's environmental policy, their implementation, monitoring, and evaluation in the context of environmental policy, objectives, targets, and legal requirements, and then acting to ensure steady improvement of the environmental management system. This approach is focused on a continuous process of enhancement and adjustment, ensuring that all aspects of an organization's operations are regularly analysed, refined, and adapted to changing environmental requirements (Mosgaard, Bundgaard, & Kristensen, 2022). In this way, the ISO 14001 standard promotes a continuous improvement model that helps organizations minimize their negative impact on the environment at the local and global levels, by striving for sustainability and conservation of natural resources.

Also crucial in the context of environmental management systems is the EMAS regulation, which was introduced by the European Commission as a binding legal act that came into force in April 1995. The main objective of EMAS is to create a green management system in organizations, based on continuous improvements in production processes and management techniques (Martins & Fonseca, 2018). To achieve this goal, companies are encouraged to create and implement alternative environmental strategies and programmes, and regularly evaluate their own efforts as well as other activities (Testa, Iraldo, & Daddi, 2018). EMAS precisely defines the responsibilities of an organization, starting with an initial environmental review, which helps identify areas for improvement and set environmental objectives, as well as monitor the transparency of its activities (Szyszka & Matuszak-Flejszman, 2017). EMAS imposes on organizations the obligation to monitor the activities of their suppliers and subcontractors so as to ensure that their production and services are based on sustainable practices.

One important aspect of EMAS is the involvement of employees at every level of the organization. All employees should be aware of their environmental responsibility and actively participate in achieving the company's environmental goals (Díaz de Junguitu & Allur, 2019). This system also requires companies to analyse the environmental impact of all their activities, processes, and products, as well as monitor and evaluate any environmental initiatives that are already in place.

Another central component of EMAS is the requirement that companies publish an environmental statement. Such a public declaration provides information about a company's environmental activities, their impact on the environment, as well as the progress made towards achieving environmental goals. It is a key element in a company's communications with its stakeholders, helping build trust and transparency in the company's environmental activities (Ociepa-Kubicka, Deska, & Ociepa, 2021).

In the face of growing pressure on companies to publish information regarding their environmental, social, and corporate governance, one tool of increasing importance is corporate sustainability reporting, informing stakeholders of the company's environmental impact (Ioannou & Serafeim, 2017). The Corporate Sustainability Reporting Directive (CSRD) aims to create the conditions for sustainability reporting so that it becomes as important a pillar as financial reporting. This is to reduce the phenomenon of greenwashing (Demarigny, 2023). To this end, the CSRD seeks to facilitate investment choices in sustainable resources by improving the assessment, comparability, and transparency of sustainability information (Baumüller & Grbenic, 2021).

The directive adopted by the European Parliament replaced the existing rules regulating non-financial disclosures (Non-Financial Reporting Directive), which were largely considered insufficient (Dolmans et al., 2021). The CSRD introduces more detailed reporting requirements regarding the environmental, human rights, and social impact of companies, based on common criteria in line with the European Union's climate goals (Dinh, Husmann, & Melloni, 2023). It increases the profile of environmental, social, and governance reporting, giving investors access to comparable and reliable data that are crucial for assessing the sustainability of enterprises and their impact on the community and the environment (Próchniak & Płoska, 2022).

A major problem affecting non-financial reporting to date has been the relatively limited comparability of corporate reports and the data presented in them. This information is often presented in the form of separate reports or as part of a statement included in the management report. Companies may use different reporting standards, both international and domestic, which further complicates the situation (Arvidsson & Dumay, 2022). Practice shows that there is also considerable freedom when it comes to the publication of such information and the level and scope of the detail. In response to these challenges, sustainability reporting in accordance with the CSRD is to be based on the new, uniform European Sustainability Reporting Standards (ESRS), the aim of which is to standardize and increase the transparency of reports, thus making them easier to compare and evaluate.

In 2022, the European Financial Reporting Advisory Group (EFRAG) proposed an initial set of draft ESRS in the following form (European Financial Reporting Advisory Group [EFRAG], 2022):

- **ESRS 1 General Requirements** – Outlining the fundamental requirements for sustainability reporting.
- **ESRS 2 General Disclosures** – Presenting the general disclosures that organizations need to include in their sustainability reports.

- **ESRS E1 Climate Change** – Focusing on reports of the impacts of climate change, and strategies related to climate change.
- **ESRS E2 Pollution** – Dealing with disclosures on pollution and how the latter is managed by an organization.
- **ESRS E3 Water and Marine Resources** – Dealing with the use and conservation of water and marine resources.
- **ESRS E4 Biodiversity and Ecosystems** – Covering various aspects of biodiversity and its impact on ecosystems.
- **ESRS E5 Resources and Circular Economy** – Focusing on the use of resources and practices connected with the CE.
- **ESRS S1 Own Workforce** – Reporting on matters relating to an organization's own workforce.
- **ESRS S2 Workers in the Value Chain** – Covering disclosures about workers in the value chain of an organization.
- **ESRS S3 Affected Communities** – Concentrating on the impact of an organization's operations on communities.
- **ESRS S4 Customers and End-Users** – Addressing various issues connected with customers and end-users of an organization's products or services.
- **ESRS G1 Business Conduct** – Dealing with an organization's business conduct and ethics.

The aim of these standards is to ensure greater transparency and comparability of sustainability reporting across Europe (EFRAG, 2022).

1.5   Conclusions

Developing successful entrepreneurial initiatives for a sustainable future is vital if organizations wish to take advantage of the opportunities that sustainability and entrepreneurship bring. According to research by Kushwaha and Kumar Sharma (2017), important motivators for sustainable entrepreneurship include environmental awareness, market opportunities, autonomy in business, livelihoods, and personal passion – these are regarded as "intrinsic" environmental factors. At the same time, external factors such as ecology-related marketing strategies, government involvement, and ecological culture play an important role. Research conducted by O'Neil and Ucbasaran (2016) highlights the importance of legitimacy for stakeholder engagement and success. However, sometimes achieving legitimacy may require trade-offs from entrepreneurs when it comes to their original motives. Despite efforts to promote sustainability through such business strategies as "greening businesses" and "triple bottom line", major global challenges, such as climate change, species extinction, and meeting basic human needs, remain. However, many expect a "greening" of the economy, based primarily on the potential for innovation and entrepreneurship (Sołtysik et al., 2024).

Integrating sustainability into business operations brings a number of benefits for entrepreneurs and their companies, which have been clearly documented in scientific research (Terán-Yépez et al., 2020), such as:

- Increased efficiency and reduced costs: implementing sustainable practices, such as minimizing waste, efficient energy management, and choosing environmentally friendly materials, often results in lower operating costs and increased overall efficiency.
- Enhanced brand image: engaging in sustainable practices can significantly improve a company's reputation, which translates into greater customer loyalty. In an era of growing consumer awareness, with customers preferring brands that reflect their own values, sustainable operations can make a company stand out from the competition.
- Attracting and retaining talent: sustainability is becoming an increasingly important priority for employees. Companies that operate sustainably are more likely to attract and retain top talent, especially those who strive to bring about positive social and environmental change.
- Strengthening stakeholder relationships: in light of the growing awareness of the environmental and social impact of business operations, a commitment to sustainability can strengthen relationships with key stakeholders, including customers, suppliers, and the community. Such an approach demonstrates a genuine commitment to their well-being.
- Long-term business viability: it is crucial that an organization take into account the long-term social and environmental consequences of business operations so as to ensure the sustainable profitability and success of a business.

Integrating sustainable practices into entrepreneurship, while beneficial, involves a number of significant challenges (Olateju, Danmola, & Aminu, 2020):

- Considerable upfront investment: switching to sustainable practices, for example, using eco-friendly materials, minimizing waste, and optimizing energy consumption, often requires significant upfront investment. Such a financial imperative can pose a significant obstacle, especially for new or smaller ventures.
- Opposition to organizational change: adapting established business processes and practices to sustainability requirements may be met with resistance, both from employees and from management and other stakeholders who are committed to traditional ways of doing things.
- Gaps in knowledge and experience: many entrepreneurs may lack the necessary knowledge or experience to successfully implement sustainable practices. This skills gap can make it difficult to initiate or effectively implement sustainability strategies.

- Resource constraints: entrepreneurs, especially those with limited resources, may find it difficult to invest in the tools and technologies necessary for sustainable development, such as energy-efficient systems or renewable energy sources.
- Competitive pressure: in some industries, adopting sustainable practices can temporarily put a company at a competitive disadvantage compared to rivals that don't make the same sustainability commitments.

Many entrepreneurs have successfully overcome these hurdles by incorporating sustainable approaches into their operations, resulting in long-term benefits. By consciously recognizing and strategically counteracting these challenges, entrepreneurs have a chance to prepare their companies for sustainable success and bolster their resilience to changing market conditions. This commitment to sustainability many not only give companies a competitive advantage but also contribute to long-term growth and stability.

Acknowledgement

The publication/article presents the result of the Project no 070/ZJE/2024/POT financed from the subsidy granted to the Krakow University of Economics.

References

Ahmed, A., Mathrani, S. and Jayamaha, N. (2024). An integrated lean and ISO 14001 framework for environmental performance: an assessment of New Zealand meat industry, *International Journal of Lean Six Sigma, 15*(3), 567–587. https://doi.org/10.1108/IJLSS-05-2021-0100

Al-Jayyousi, O., Amin, H., Al-Saudi, H. A., Aljassas, A., & Tok, E. (2023). Mission-oriented innovation policy for sustainable development: A systematic literature review. *Sustainability, 15*(17), 13101. https://doi.org/10.3390/su151713101

Arvidsson, S., & Dumay, J. (2022). Corporate ESG reporting quantity, quality and performance: Where to now for environmental policy and practice? *Business Strategy and the Environment,* 31(3), 1091–1110.

Baumgartner, R. J. (2014). Managing corporate sustainability and CSR: A conceptual framework combining values, strategies and instruments contributing to sustainable development. *Corporate Social Responsibility and Environmental Management,* 21, 258–271.

Baumüller, J., & Grbenic, S. (2021). Moving from non-financial to sustainability reporting: analyzing the EU Commission's proposal for a Corporate Sustainability Reporting Directive (CSRD). *Facta Universitatis, Series: Economics and Organization, 18*(4), 369–381. https://doi.org/10.22190/FUEO210817026B.

Bravi, L., Santos, G., Pagano, A., & Murmura, F. (2020). Environmental management system according to ISO 14001: 2015 as a driver to sustainable development. *Corporate Social Responsibility and Environmental Management,* 27(6), 2599–2614.

Caiado, R. G. G., de Freitas Dias, R., Mattos, L. V., Quelhas, O. L. G., & Leal Filho, W. (2017). Towards sustainable development through the perspective of eco-efficiency – A systematic literature review. *Journal of Cleaner Production,* 165, 890–904. https://doi.org/10.1016/j.jclepro.2017.07.166

Cecchin, A., Salomone, R., Deutz, P., Raggi, A., & Cutaia, L. (2021). What is in a name? The rising star of the circular economy as a resource-related concept for sustainable development. *Circular Economy and Sustainability*. https://doi.org/10.1007/s43615-021-00021-4

Chancel, L., Piketty, T., Saez, E., & Zucman, G. (Eds.) (2022). *World inequality report 2022*. Harvard University Press.

Cohen, B. (2006). Sustainable valley entrepreneurial ecosystems. *Business Strategy and the Environment*, 15(1), 1–14. https://doi.org/10.1002/bse.428

Cohen, B., & Winn, M. I. (2007). Market imperfections, opportunity and sustainable entrepreneurship. *Journal of Business Venturing*, 22(1), 29–49.

Dangelico, R. M. (2016). Green product innovation: Where we are and where we are going. *Business Strategy and the Environment*, 25(8), 560–576. https://doi.org/10.1002/bse.1886

de Oliveira Neves, F., Salgado, E. G., & Beijo, L. A. (2017). Analysis of the environmental management system based on ISO 14001 on the American continent. *Journal of Environmental Management*, 199, 251–262.

Demarigny, F. (2023). Sustainability information and financial market efficiency. *AEFR Debate Paper* (1).

Dey, P. K., Petridis, N. E., Petridis, K., Malesios, C., Nixon, J. D., & Ghosh, S. K. (2018). Environmental management and corporate social responsibility practices of small and medium-sized enterprises. *Journal of Cleaner Production*, 195, 687–702. https://doi.org/10.1016/j.jclepro.2018.05.201

Díaz de Junguitu, A., & Allur, E. (2019). The adoption of environmental management systems based on ISO 14001, EMAS, and alternative models for SMEs: A qualitative empirical study. *Sustainability*, 11(24), 7015.

Dinh, T., Husmann, A., & Melloni, G. (2023). Corporate sustainability reporting in Europe: A scoping review. *Accounting in Europe*, 20(1), 1–29.

Dolmans, M., Bourguignon, G., Assereto, C. C., & Dictus, T. (2021). The corporate sustainability reporting directive: From "non-financial" to "sustainability" reporting. Cleary Gottlieb Steen & Hamilton LLP, 1–7. Memorandum: https://www.clearygottlieb.com/-/media/files/alert-memos-2021/the-corporate-sustainability-reporting-directive.pdf

European Financial Reporting Advisory Group (EFRAG) (2022). First set of draft European sustainability reporting standards (ESRS). Retrieved from https://www.efrag.org/lab6

Falcone, P. M., Tani, A., Tartiu, V. E., & Imbriani, C. (2020). Towards a sustainable forest-based bioeconomy in Italy: Findings from a SWOT analysis. *Forest Policy and Economics*, 110, 101910. https://doi.org/10.1016/j.forpol.2019.04.014

Farny, S., & Binder, J. K. (2021). Sustainable entrepreneurship. In Léo-Paul Dana (Ed.), *World Encyclopedia of entrepreneurship* (pp. 605–611). Edward Elgar Publishing

Ferrón Vílchez, V., & Darnall, N. (2016). Two better than one: The link between management systems and business performance. *Business Strategy and the Environment*, 25, 221–240.

González-Ruiz, J. D., Botero-Botero, S., & Duque-Grisales, E. (2018). Financial eco-innovation as a mechanism for fostering the development of sustainable infrastructure systems. *Sustainability (Switzerland)*. https://doi.org/10.3390/su10124463

Gupta, S. (2020). Understanding the feasibility and value of grassroots innovation. *Journal of the Academy of Marketing Science*, 48, 941–965. https://doi.org/10.1007/s11747-019-00639-9

Kushwaha, G. S., & Kumar Sharma, N. (2017). Factors influencing young entrepreneurial aspirant's insight towards sustainable entrepreneurship. *Interdisciplinary Journal of Management Studies (Formerly known as Iranian Journal of Management Studies)*, 10(2), 435–466. https://doi.org/10.22059/ijms.2017.224885.672467.

Haas, W., Krausmann, F., Wiedenhofer, D., & Heinz, M. (2015). How circular is the global economy?: An assessment of material flows, waste production, and recycling in the European Union and the world in 2005. *Journal of Industrial Ecology*, 19(5), 765–777. https://doi.org/10.1111/jiec.12244

Halberstadt, J., Schwab, A.-K., & Kraus, S. (2024). Cleaning the window of opportunity: Towards a typology of sustainability entrepreneurs. *Journal of Business Research*, 171, 114386. https://doi.org/10.1016/j.jbusres.2023.114386

Heras-Saizarbitoria, I., Urbieta, L., & Boiral, O. (2022). Organizations' engagement with sustainable development goals: From cherry-picking to SDG-washing? *Corporate Social Responsibility and Environmental Management*, 29(2), 316–328.

Hristov, I., Chirico, A., & Ranalli, F. (2022). Corporate strategies oriented towards sustainable governance: Advantages, managerial practices and main challenges. *Journal of Management and Governance*, 26(1), 75–97.

Ikram, M., Zhou, P., Shah, S. A. A., & Liu, G. Q. (2019). Do environmental management systems help improve corporate sustainable development? Evidence from manufacturing companies in Pakistan. *Journal of Cleaner Production*, 226, 628–641. https://doi.org/10.1016/j.jclepro.2019.03.265

Ioannou, I., & Serafeim, G. (2017). The consequences of mandatory corporate sustainability reporting. Harvard Business School Research Working Paper (pp. 11–100).

Jansen, L. (2003). The challenge of sustainable development. *Journal of Cleaner Production*, 11(3), 231–245.

Johnson, M. P., & Schaltegger, S. (2020). Entrepreneurship for sustainable development: A review and multilevel causal mechanism framework. *Entrepreneurship Theory and Practice*, 44(6), 1141–1173. https://doi.org/10.1177/1042258719885368

Ketprapakorn, N., & Kantabutra, S. (2019). Sustainable social enterprise model: Relationships and consequences. *Sustainability*, 11(14), 3772. https://doi.org/10.3390/su11143772

Kickul, J., & Lyons, T. S. (2020). *Understanding social entrepreneurship: The relentless pursuit of mission in an ever changing world*. Routledge.

Kiefer, C. P., Del Río González, P., & Carrillo-Hermosilla, J. (2019). Drivers and barriers of eco-innovation types for sustainable transitions: A quantitative perspective. *Business Strategy and the Environment*, 28(1), 155–172. https://doi.org/10.1002/bse.2246

Koe, W. L., Omar, R., & Sa'ari, J. R. (2015). Factors influencing propensity to sustainable entrepreneurship of SMEs in Malaysia. *Procedia-Social and Behavioral Sciences*, 172, 570–577.

Korhonen, J., Honkasalo, A., & Seppälä, J. (2018). Circular economy: The concept and its limitations. *Ecological Economics*. https://doi.org/10.1016/j.ecolecon.2017.06.041

Kuckertz, A. (2020). Bioeconomy transformation strategies worldwide require stronger focus on entrepreneurship. *Sustainability*, 12(7), 2911. https://doi.org/10.3390/su12072911

Laner, D., & Rechberger, H. (2009). Quantitative evaluation of waste prevention on the level of small and medium sized enterprises (SMEs). *Waste Management*, 29(2), 606–613. https://doi.org/10.1016/j.wasman.2008.05.007

Lee, S. M., Noh, Y., Choi, D., & Rha, J. S. (2017). Environmental policy performances for sustainable development: From the perspective of ISO 14001 certification. *Corporate Social Responsibility and Environmental Management*, 24(2), 108–120.

Liao, Y. (2022). Sustainable leadership: A literature review and prospects for future research. *Frontiers in Psychology*, 13, 1045570.

Lucendo-Monedero, Á. L., Ruiz-Rodríguez, F., & González-Relaño, R. (2023). The information society and socio-economic sustainability in European regions. Spatio-temporal changes between 2011 and 2020. *Technology in Society*, 75, 102337.

Lüdeke-Freund, F., Gold, S., & Bocken, N. M. P. (2019). A review and typology of circular economy business model patterns. *Journal of Industrial Ecology*, 23(1), 36–61. https://doi.org/10.1111/jiec.12763

Ma, S., Wen, Z., Chen, J., & Wen, Z. (2014). Mode of circular economy in China's iron and steel industry: A case study in Wu'an city. *Journal of Cleaner Production*, 64, 505–512. https://doi.org/10.1016/j.jclepro.2013.10.008

Malhotra, S., & Kiran, R. (2023). Examining the relationship between entrepreneurial perceived behaviour, intentions, and competencies as catalysts for sustainable growth: An Indian perspective. *Sustainability*, 15(8), 6617.

Martins, F., & Fonseca, L. (2018). Comparison between eco-management and audit scheme and ISO 14001: 2015. *Energy Procedia*, 153, 450–454.

Merli, R., Preziosi, M., & Acampora, A. (2018). How do scholars approach the circular economy? A systematic literature review. *Journal of Cleaner Production*, 178, 703–722. https://doi.org/10.1016/j.jclepro.2017.12.112

Millar, N., McLaughlin, E., & Börger, T. (2019). The circular economy: Swings and roundabouts? *Ecological Economics*, 158, 11–19. https://doi.org/10.1016/j.ecolecon.2018.12.012

Mosgaard, M. A., Bundgaard, A. M., & Kristensen, H. S. (2022). ISO 14001 practices– A study of environmental objectives in Danish organizations. *Journal of Cleaner Production*, 331, 129799.

Muktiono, E., & Soediantono, D. (2022). Literature review of ISO 14001 environmental management system benefits and proposed applications in the defense industries. *Journal of Industrial Engineering & Management Research*, 3(2), 1–12.

Muñoz, P., & Cohen, B. (2018). Sustainable entrepreneurship research: Taking stock and looking ahead. *Business Strategy and the Environment*, 27(3), 300–322. https://doi.org/10.1002/bse.2000

Naciti, V., Cesaroni, F., & Pulejo, L. (2021). Corporate governance and sustainability: A review of the existing literature. *Journal of Management and Governance*, 26(1), 55–74.

Niemann, C. C., Dickel, P., & Eckardt, G. (2020). The interplay of corporate entrepreneurship, environmental orientation, and performance in clean-tech firms—A double-edged sword. *Business Strategy and the Environment*, 29(1), 180–196. https://doi.org/10.1002/bse.2357

Nitkiewicz, T. (2021). Facing circularity and sustainability challenge in agricultural sector – LCA approach towards assessment of its consequences. In *Current trends in quality science – Consumer behavior, logistic, product management* (ed.) Śmigielska H. (pp. 246–260), Sieć Badawcza Łukasiewicz – Instytut Technologii Eksploatacji. https://www.itee.lukasiewicz.gov.pl/images/stories/exlibris/SMIGIELSKA_Hanna.pdf

Nowicki, P., Ćwiklicki, M., Kafel, P., Niezgoda, J., & Wojnarowska, M. (2023). The circular economy and its benefits for pro-environmental companies. *Business Strategy and the Environment*. https://doi.org/10.1002/bse.3382

O'Neil, I., & Ucbasaran, D. (2016). Balancing "what matters to me" with "what matters to them": Exploring the legitimation process of environmental entrepreneurs. *Journal of Business Venturing*, 31(2), 133–152.

Ociepa-Kubicka, A., Deska, I., & Ociepa, E. (2021). Organizations towards the evaluation of environmental management tools ISO 14001 and EMAS. *Energies*, 14(16), 4870. https://doi.org/10.3390/en14164870

Olateju, A. O., Danmola, R. A., & Aminu, A. W. (2020). Sustainable entrepreneurship and sustainable development in Nigeria: Prospects and challenges. *International Journal of Research and Innovation in Social Science*, 4(11), 372.

Özkan, P., & Karataş Yücel, E. (2020). From linear economy to circular economy. https://doi.org/10.4018/978-1-7998-5116-5.ch004

Pacana, A. (2018). Zarządzanie środowiskowe zgodne z ISO 14001:2015. Oficyna Wydawnicza Politechniki Rzeszowskiej, Rzeszów.

Palmer, K., Oates, W. E., & Portney, P. R. (1995). Tightening environmental standards: The benefit-cost or the no-cost paradigm? *Journal of Economic Perspectives*, 9(4), 119–132. https://doi.org/10.1257/jcp.9.4.119

Pandey, N., Andres, C., & Kumar, S. (2023). Mapping the corporate governance scholarship: Current state and future directions. *Corporate Governance: An International Review*, 31(1), 127–160.

Parrish, B. D., & Foxon, T. J. (2006). Sustainability entrepreneurship and equitable transitions to a low-carbon economy. *Greener Management International*, 2006(55), 47–62.

Patzelt, H., & Shepherd, D. A. (2011). Recognizing opportunities for sustainable development. *Entrepreneurship Theory and Practice*, 35(4), 631–652.

Pereira, J., Rodrigues, R. G., & Veiga, P. M. (2024). Entrepreneurship among social workers: Implications for the sustainable development goals. *Sustainability*, 16(3), 996. https://doi.org/10.3390/su16030996

Porter, M. E., & Linde, C. van der. (1995). Toward a new conception of the environment-competitiveness relationship. *Journal of Economic Perspectives*, 9(4), 97–118. https://doi.org/10.1257/jep.9.4.97

Próchniak, J., & Płoska, R. (2022). WIG-20 Warsaw Stock Exchange companies: Are they ready for governance matters disclosures based on EU sustainable reporting standards? Annales Universitatis Mariae Curie-Skłodowska, *Sectio H Oeconomia*, 56(5), 227–246, https://doi.org/10.17951/h.2022.56.5.227-246.

Psomas, E. L., Fotopoulos, C. V., & Kafetzopoulos, D. P. (2011). Motives, difficulties and benefits in implementing the ISO 14001 environmental management system. *Management of Environmental Quality: An International Journal*, 22(4), 502–521.

Ratten, V., Jones, P., Braga, V., & Marques, C. S. (2019). *Sustainable entrepreneurship: The role of collaboration in the global economy* (pp. 1–7). Springer International Publishing.

Ruiz-Real, J. L., Uribe-Toril, J., Valenciano, J. D. P., & Gázquez-Abad, J. C. (2018). Worldwide research on circular economy and environment: A bibliometric analysis. *International Journal of Environmental Research and Public Health*, 15(12). https://doi.org/10.3390/ijerph15122699

Ruiz-Mercader, J., Meroño-Cerdan, A. L., & Sabater-Sánchez, R. (2006). Information technology and learning: Their relationship and impact on organisational performance in small businesses. *International Journal of Information Management*, 26(1), 16–29. https://doi.org/10.1016/j.ijinfomgt.2005.10.003

Salmivaara, V., & Kibler, E. (2020). "Rhetoric mix" of argumentations: How policy rhetoric conveys meaning of entrepreneurship for sustainable development. *Entrepreneurship Theory and Practice*, 44(4), 700–732. https://doi.org/10.1177/1042258719845345

Sarango-Lalangui, P., Santos, J. L. S., & Hormiga, E. (2018). The development of sustainable entrepreneurship research field. *Sustainability*, 10(6), 2005. https://doi.org/10.3390/su10062005

Seroka-Stolka, O. (2023). Towards sustainability: An environmental strategy choice, environmental performance, and the moderating role of stakeholder pressure. *Business Strategy and the Environment*, 32(8), 5992–6007. https://doi.org/10.1002/bse.3469

Smol, M., & Kulczycka, J. (2019). Towards innovations development in the European raw material sector by evolution of the knowledge triangle. *Resources Policy*, 62, 453–462. https://doi.org/10.1016/j.resourpol.2019.04.006

Sołtysik, M., Urbaniec, M., & Wojnarowska, M. (2019). Innovation for sustainable entrepreneurship: Empirical evidence from the bioeconomy sector in Poland. *Administrative Sciences*, 9(3), 50. https://doi.org/10.3390/admsci9030050

Sołtysik, M., Wojnarowska, M., Urbaniec, M., Zabkar, V., Ćwiklicki, M., & Varese E. (2024). *Sustainable business in the era of digital transformation: Strategic and entrepreneurial perspectives.* Abingdon; New York: Routledge.

Spiegler, A. B., & Halberstadt, J. (2018). SHEstainability: How relationship networks influence the idea generation in opportunity recognition process by female social entrepreneurs. *International Journal of Entrepreneurial Venturing*, 10(2), 202–235.

Steger, U. (2000). Environmental management systems: Empirical evidence and further perspectives. *European Management Journal*, 18(1), 23–37. https://doi.org/10.1016/s0263-2373(99)00066-3

Štrukelj, T., Dankova, P., & Hrast, N. (2023). Strategic transition to sustainability: A cybernetic model. *Sustainability*, 15(22), 15948. https://doi.org/10.3390/su152215948

Suriyankietkaew, S., & Kungwanpongpun, P. (2022). Strategic leadership and management factors driving sustainability in health-care organizations in Thailand. *Journal of Health Organization and Management*, 36(4), 448–468.

Szyszka, B., & Matuszak-Flejszman, A. (2017). EMAS in Poland: Performance, effectiveness, and future perspectives. *Polish Journal of Environmental Studies*, 26(2), 809–817. https://doi.org/10.15244/pjoes/65544

Terán-Yépez, E., Marín-Carrillo, G. M., del Pilar Casado-Belmonte, M., & de las Mercedes Capobianco-Uriarte, M. (2020). Sustainable entrepreneurship: Review of its evolution and new trends. *Journal of Cleaner Production*, 252, 119742.

Testa, F., Iraldo, F., & Daddi, T. (2018). The effectiveness of EMAS as a management tool: A key role for the internalization of environmental practices. *Organization & Environment*, 31(1), 48–69.

Tudoran, V. I., Marin, C. A., & Condrea, E. (2020). Corrections and corrective actions–instruments for the continuous improvement of a management system. *Strategica*, 275–287.

van Berkel, R., & Kortman, J. J. G. M. (1993). Waste prevention in small and medium sized enterprises. *Journal of Cleaner Production*, 1(1), 21–28. https://doi.org/10.1016/0959-6526(93)90029-B

Volkmann, C., Fichter, K., Klofsten, M., & Audretsch, D. B. (2021). Sustainable entrepreneurial ecosystems: An emerging field of research. *Small Business Economics*, 56(3), 1047–1055. https://doi.org/10.1007/s11187-019-00253-7

Wagner, M. (2009). Innovation and competitive advantages from the integration of strategic aspects with social and environmental management in European firms. *Business Strategy and the Environment*, 18(5), 291–306. https://doi.org/10.1002/bse.585

Wang, Z., & Sarkis, J. (2017). Corporate social responsibility governance, outcomes, and financial performance. *Journal of Cleaner Production*, 162, 1607–1616.

Wohlgemuth, V., & Wenzel, M. (2016). Dynamic capabilities and routinization. *Journal of Business Research*, 69(5), 1944–1948. https://doi.org/10.1016/j.jbusres.2015.10.085

Wojnarowska, M., Sołtysik, M., & Ingrao, C. (2022). Characteristics of sustainable products. In *Sustainable products in the circular economy* (pp. 1–17). Routledge. https://doi.org/10.4324/9781003179788-1

Zgrzywa-Ziemak, A. (2019). Model zrównoważenia przedsiębiorstwa. *Oficyna Wydawnicza Politechniki Wrocławskiej*. ISBN: 978-83-7493-067-3.

2

PRODUCT TRANSFORMATION

From conventional to sustainable products

Mateusz Wygoda and Carlo Ingrao

2.1   Introduction

The current trends in the consumption of abiotic elements and fossil fuels, combined with the problem of global environmental pollution, highlight the need for environmentally friendly and user-friendly products. These trends largely stem from overproduction and the lack of well-thought-out purchasing decisions (Vasile Dinu et al., 2012). A number of purchased products often fail to fulfil their purpose or are used inappropriately due, for example, to a failure to understand the instructions of use, inadequate knowledge of a product, low-quality materials, or manufacturing defects. Other factors may include poor maintenance, improper usage, design flaws, or unrealistic expectations regarding a product's functionality (Fook & McNeill, 2020). These problems need to be fixed by striving in a responsible manner to manufacture products that are functional, user-friendly, and sustainable. The transformation of a product from a conventional to a sustainable commodity is a comprehensive, multi-action process that requires the involvement of producers, consumers, and other stakeholders (Roome & Anastasiou, 2002). Doing so can help reduce the negative impact of products on the environment and contribute to building a more sustainable future.

The transformation of products plays a crucial role in fostering sustainable development (Țigan et al., 2021; Vasile Dinu et al., 2012). In the economic domain, they enhance resource efficiency management by minimizing waste, promoting reuse and recycling, and incorporating ecological design practices. Furthermore, manufacturers of sustainable products actively promote their own efforts to reduce energy consumption, and prioritize the utilization of local and renewable resources to diminish the water footprint. Social and cultural values are instrumental in revitalizing traditional varieties, reintroducing authentic recipes and technologies,

DOI: 10.4324/9781032710693-3

fostering social responsibility in consumption, and nurturing demands for innovative products, as well as European ecolabel products and services (Țigan et al., 2021). Recognizing and responsibly harnessing novel resources is vital for diversification, while simultaneously opening up the potential for the natural and sustainable reestablishment of a dedicated eco-friendly industry (Ferraris et al., 2021). All organizations, irrespective of the economic sectors in which they operate, must undertake environmentally friendly initiatives through innovative mechanisms. This involves taking into account the current situation and future prospects regarding natural resources, their key characteristics, and their limited availability (Calabrese et al., 2021). Due to the above-described problems, actions should be taken aimed at pursuing and implementing responsible product sustainability. It has been proven in the literature that this can be done by taking into account the three dimensions of sustainability: economic, environmental, and social and cultural (Țigan et al., 2021).

In terms of the economic area, the following factors can be identified for waste reduction (Yadav et al., 2021): reuse, recycling, environmentally conscious design (eco-design) (Chia et al., 2020), and the cultivation of crop varieties tailored to local ecological conditions (Klaus & Kiehl, 2021; Wan et al., 2020). In the second environmental area, various factors contribute to sustainability, including lower greenhouse gas emissions (Rahman et al., 2021), energy conservation (Rosa et al., 2021; Xu & Szmerekovsky, 2017), decreased toxicity (Dastan et al., 2019), the utilization of local renewable resources (Das et al., 2020), diminished water usage (Palhares et al., 2021a, 2021b; Rahman et al., 2021), and waste management practices, such as composting (Lopes et al., 2021; H. B. Sharma et al., 2021). In the case of social and cultural aspects, key factors include preserving traditional cultivars, recipes, and technologies (Bernardi et al., 2021), fostering social responsibility and sustainable consumption patterns (Yusoff et al., 2019), promoting innovation (Alqalami et al., 2020), ensuring food safety and health (Ferraris et al., 2021), and prioritizing education (M. Corrêa et al., 2020).

Education emerges as a pivotal element driving forward sustainable practices. It serves as a primary indicator of future environmental planning. Therefore, there is a pressing need to revamp education systems and facilitate responsible teaching that steers individuals towards sustainability (M. Corrêa et al., 2020; Țigan et al., 2021).

Furthermore, the European Union (EU) treaties address the issue of the current "take-make-dispose" economic model (European Commission – About Sustainable Products, 2024). This type of activity depletes our resources, contaminates our environment, and harms both the biodiversity and the climate. Furthermore, it reinforces Europe's dependence on resources from external sources. In response to these challenges, the EU is seeking to bring about a transition towards a circular economy model, putting more emphasis on the production of more sustainable products (European Commission – About Sustainable Products, 2024; European Commission – Circular Economy Action Plan, 2024).

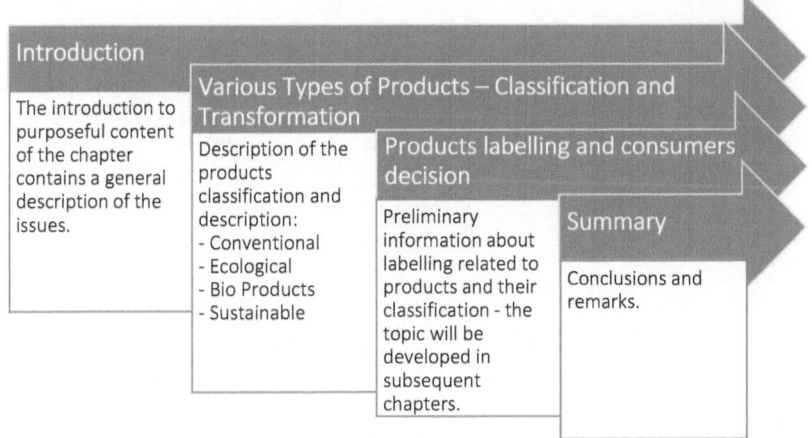

FIGURE 2.1   Chart of chapter structure.

Improving management of the environmental impact of products throughout their life cycle and extending their usage will help develop more sustainable, circular, and resource-efficient products within the EU. Integrating sustainability measures in products such as electronics, furniture, and textiles will strengthen the resilience of the EU economy (European Commission – About Sustainable Products, 2024; European Commission – Circular Econo.my Action Plan, 2024; Sengstschmid et al., 2011).

The aim of this chapter is to present information about different types of products. This kind of information can be applied in the form of specific knowledge about products addressed to society and other economic entities. The structure of this chapter is presented in Figure 2.1.

2.2   Various types of products – classification and transformation

In today's globalized world, the market offers a variety of products that are designed to meet consumers' needs, and which, over time, undergo transformation and be updated. Generally, the term "product transformation" can refer to various contexts depending on the industry (Zhang et al., 2023). Overall, however, this term is associated with the processes of changing, improving, or modifying products to bring them into line with new requirements, market trends, or customer needs (Markeset & Kumar, 2005). Within a specified scope, steps need to be taken to turn conventional products into sustainable versions. One such measure involves introducing innovative elements into the manufacturing process, with the aim of increasing their value, efficiency, and attractiveness for customers, or adapting them to the needs of new technologies (Herrmann et al., 2007). Other necessary actions may include making improvement-oriented modifications and updates to the technologies and software currently used in product manufacturing systems.

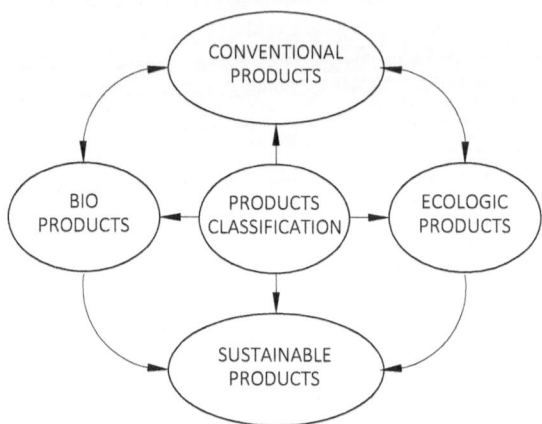

FIGURE 2.2   General classification of product types.

This approach may entail making adjustments to production processes so as to enhance the functionality, safety, and overall quality of products, while limiting any increase in their market price (Al-Ashaab et al., 2013). Other possible measures include adapting a product to market needs to better meet customer expectations (Stolz et al., 2011), reflecting changing consumer trends, and competing with other products on the market (Garekwe et al., 2024).

In the business domain, the product transformation process may require adapting to the conditions of digital transformation, which might necessitate integrating and implementing new technologies such as artificial intelligence or data analytics (Coronado-Medina et al., 2020).

However, the current literature includes numerous informal classifications of products (Wojnarowska et al., 2022). In light of this fact, the authors devised the classification presented in Figure 2.2.

The above classification not only categorizes products based on industry types but also encompasses factors that take into account the entire process of their production, distribution, and utilization by end-users. The diagram presents a very broad and complex scope of knowledge that does not fit within the scope of a single available market product or service. Transforming a product from conventional to sustainable is a complex process that requires the involvement of producers, consumers, and other stakeholders. These actions can help reduce the negative impact of products on the environment and contribute to a more sustainable future. Therefore, in the following subsections, an attempt is made to provide a synthetic description of the aforementioned products and their classification.

2.2.1   Conventional products

Conventional products, also known as classical products (Brzostek-Kasprzak et al., 2013), encompass items typically derived from production methods geared towards maximizing quantity while minimizing costs. These products are often characterized by their affordability, achieved through efficient manufacturing processes. Lower production costs translate into competitive pricing, rendering conventional products very widely available and sought-after in larger market contexts. Such production systems are essentially based on traditional and widespread technologies that, however, may have a negative impact on the environment and human health (Markeset & Kumar, 2005; Meier et al., 2015). One example of this is intensive farming systems that, to maximize yields and incomes, require the application of chemical fertilizers and pesticides, as well as high fuel consumption levels for soil treatment due to the use of synthetic chemicals, pesticides, and excessive exploitation of natural resources (Meier et al., 2015). For example, in the case of food-focused agricultural production, besides the use of chemical agents, the genetic modification of plants is also permitted (Malinowski et al., 2019). Similarly, in other industries, such as manufacturing and construction, there is a reliance on resource-intensive processes and materials, which often leads to environmental degradation and unsustainable practices (Okogwu et al., 2023).

2.2.2 Ecological products

Previous studies analyzing the relationship between sustainable development and products have mainly focused on eco-friendly products, that is, those that have a less negative impact on the environment than the conventional products (Bhardwaj et al., 2020; Biswas & Roy, 2015; Nuryakin & Maryati, 2020; Qiu & Cardinale, 2020; Sdrolia et al., 2019; Tezer & Bodur, 2020). There are approximately 50 definitions of eco-friendly products, indicating the diversity of approaches in this regard (Sdrolia et al., 2019). As a consequence, eco-friendly products are designed with the aim of protecting or improving the environment by conserving energy and/or resources, as well as by reducing or eliminating the use of toxic substances, pollution, and waste (Ottman et al., 2006).

Research on the environmental impact of products has significantly contributed to our understanding of how companies can develop more eco-friendly products, which should result in success for a company, although this is not always the case (Hofenk et al., 2019). However, what is emphasized is that products manufactured during the production process have an impact not only on the environment but also on society, including employees, business owners, communities, and customers, throughout the entire product lifecycle. Therefore, it is important not only to optimize the impact of products on the environment but also on society as a whole (Lin et al., 2018).

In this context, eco-design and life cycle assessment (LCA) play a significant role. Eco-design involves designing products in a way that minimizes their environmental impact from the design stage by using eco-friendly materials, optimizing production processes, and ensuring long-term use and easy disposal. On the other hand, LCA enables a comprehensive analysis of the environmental impact of a product at all stages of its lifecycle, from raw material extraction, through production, distribution, and use, right up to its disposal, which makes it possible to identify areas where improvements can be made in terms of sustainable development. The concepts of eco-design and LCA are described in more details in Chapter 7. Integrating these approaches allows a holistic method for creating products that are both environmentally friendly and socially beneficial.

2.2.3 Bioproducts

Bioproducts are produced using biotic resources, including in a mix with abiotic components (Ontario, 2024). Biological resources include forestry, agricultural elements (e.g., crops and crop residues, dried distillers' grains) and biologically derived waste or other renewable bioresources. Regarding the advantages of bioproducts, the up-and-coming bio-economy offers solutions to global crises in energy, the socioeconomic sphere, and the environment (Thangaraj et al., 2018).

When it comes to energy-related benefits, it is worth noting that a responsible feedstock source reduces dependence on foreign and fossil fuel markets. This ensures greater energy security with fewer energy imports. At the beginning of the supply chain, local biomass cultivation and retrieval are utilized, while at the end there is minimal air and water contamination (Thangaraj et al., 2018).

In the realm of socioeconomic benefits, it is noteworthy that the development of new industries and products not only fosters economic growth but also provides more job opportunities, particularly in rural areas. Such growth not only stimulates local economies but also enhances the overall quality of life for residents, ultimately contributing to sustainable community development (Ontario, 2024). The environmental benefits include lower greenhouse gas emissions from the manufacture of certain bioproducts compared to petroleum-based equivalents, less harm to the environment, reduced toxicity, as well as increased biodegradability and sustainable production of renewable feedstocks (Ontario, 2024). Bioproducts are heralded as a significant opportunity for utilizing renewable resources. This highlights their potential to mitigate reliance on finite fossil fuels and instead harness sustainable, natural materials for various applications. By promoting the utilization of bioproducts, we not only reduce environmental impact but also pave the way for a more sustainable and eco-friendly approach to production and consumption (Smith et al., 2021). The current trend is towards a greater focus on feedstock as well as on research into synthesizing new biomass sources. In many cases, biomass can be transformed into bio-based products, packaging, or alternative power (Thangaraj et al., 2018). In this case, bioproducts can be divided into three categories: biochemicals, bioenergy, and biomaterials, which are presented in detail in Table 2.1.

Among bioproducts, bioplastics stand out as particularly noteworthy. They offer intriguing potential due to their ability to address environmental concerns associated with traditional plastics, such as fossil fuel dependence and non-biodegradability. Bioplastics offer a promising avenue for sustainable packaging and material solutions, showcasing innovation in the transition towards more eco-friendly alternatives (Jones et al., 2020). Overall, bioplastics are generally regarded as plastics that can be either bio-based, biodegradable, or possess both properties (European Bioplastics, 2024). Bio-based plastics are typically made from renewable resources, primarily plant biomass, and may have the same durability and properties as conventional plastics (Gerassimidou et al., 2021; Azhar et al., 2022). An up-to-date classification of bioplastics into four subgroups, together with related material examples, is presented in Figure 2.3.

TABLE 2.1 The categories of bioproducts

Bioenergy	Biochemicals	Biomaterials
Liquid fuels in the form of ethanol or biodiesel. Solid biomass used in the combustion process to generate heat and power. Gaseous fuel, such as biogas and syngas, used to generate heat and power.	Industrial – basic and specialty chemicals or resins, including paints, lubricants and solvents. Pharmaceuticals – antibodies and vaccines manufactured by genetically modified plants and natural source medicinal compounds. Biocosmetics in the form of soaps, body creams and lotions.	Bioplastics from plant oils and sugars. Biofoams and biorubber from plant oils and latex. Biocomposites manufactured from agricultural and forestry biofibres, used in the production of automotive door panels and parts.

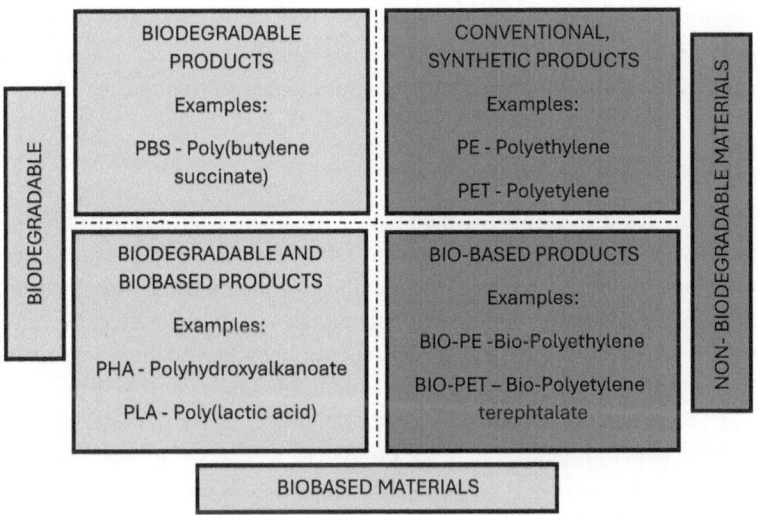

FIGURE 2.3 Examples of bioplastic products made from different materials.

Information about bioproducts should be explained based on the definitions of compostable, biodegradable, and oxo-biodegradable materials as they highlight the main differences between these bioproducts. According to Azhar et al. (2022), the following descriptions are possible:

Compostable materials – these materials break down into carbon dioxide, biomass, and water within a specific time frame in suitable composting conditions. The latter, generally, include fruit and vegetable scraps, paper tea bags, eggshells, or untreated wood chips (Composting At Home | US EPA, 2024). Compostable materials leave behind no toxic or visible residue. They can break down completely in a natural environment, providing soil with nutrient release. Compostable materials cannot be recycled (Azhar et al., 2022).

Biodegradable materials – in the presence of microorganisms, bacteria, and fungi, these materials break down into carbon dioxide, biomass, and water. If toxic elements do not break down, residue may remain in the environment. The breakdown process may be prolonged, contingent upon the disposal environment. Biodegradable materials cannot supply nutrients to the soil; however, they can be recycled (Azhar et al., 2022).

Oxo-biodegradable materials – these materials break down into smaller elements in the presence of heavy metals or other catalysts. They leave behind heavy metals or other toxins in the environment. Their complete breakdown and biodegradation still require verification. They do not provide the soil with nutrients. Oxo-biodegradable materials are recyclable (Azhar et al., 2022).

Attention should also be paid to the fact that much of the public confusion surrounding bio-based plastics stems from the use of the term "bioplastic". This term is misrepresented on the labels of plastic products. This frequently leads to waste disposal issues, whereby bioproducts are not sorted at the end of their lifecycle. This results in significant environmental issues (Gerassimidou et al., 2021).

2.2.4   Sustainable products

The literature features numerous attempts by authors to define sustainable products (Moldavska & Welo, 2017). Research into the definition of sustainable products reveals a lack of understanding of the fact that our planet itself is not a sustainable system. Only by adopting this assumption can a sustainable product be defined as follows: "a product that will have the least possible impact on the environment during its lifecycle". According to this simple definition, the lifecycle includes extraction of raw material, production, use, and final recycling (or deposition). This also includes the materials contained in the product as well as the materials or elements used for producing energy (Ljungberg, 2007).

Shuaib et al. delineate a sustainable product within the framework of sustainable development, asserting that such products yield environmental, societal, and economic benefits while safeguarding public health, welfare, and the environment throughout their entire commercial life cycle. Additionally, the authors stress that designing and manufacturing sustainable products necessitates a comprehensive approach that takes into account the economic, environmental, and social aspects of the triple bottom line. To accomplish this, attention must be paid to all stages of a product's life cycle. This holistic approach often entails adopting the 10Rs to ensure a circular material flow throughout the life cycle (Wojnarowska et al., 2022). To provide an accurate definition of what constitutes a sustainable product, the guidelines provided by EU directives (European Commission – About Sustainable Products, 2024) should be adopted as the foundational framework.

This ensures the unambiguity of both the message and the definition. Sustainable products are described as manufactured or processed in such a way as to provide environmental, social, and economic benefits (Trojanowski, 2017). Through such actions, it is possible to protect public health and the environment throughout the entire life cycle – from the initial extraction of raw resources right up until the final disposal of the product. Extending the lifetime of products and considering their environmental impact throughout their life cycle will help create more sustainable, circular, and resource-efficient products (European Commission – About Sustainable Products, 2024).

The foundational concept of a sustainable product, according to Belz and Peattie (2012) and Machindra Gurme (2022), consists of the following six characteristics:

- Customer satisfaction: ensuring products or services meet customer needs and expectations.
- Dual focus on social and ecological significance: emphasizing both the social and ecological aspects of sustainability.
- Life-cycle orientation: maintaining a product's environmentally friendly status throughout its entire life cycle.
- Significant improvements: helping solve socio-ecological problems globally by means of measurable enhancements in product performance.
- Continuous improvement: strive for ongoing enhancements in the social and environmental aspects of products.
- Benchmarking against competitors: while sustainable products may lag behind competing offers, they can serve as benchmarks for social and ecological performance.

According to the European Commission (European Commission – About Sustainable Products, 2024; European Commission – Ecodesign for Sustainable Products Regulation, 2024), the requirements for sustainable products must meet a certain number of assumptions. To move in this direction, a framework has been created to establish eco-design requirements for specific product groups. This step is aimed at achieving substantial improvements in circularity, energy performance, and other aspects of environmental sustainability. It will enable performance and information requirements to be set for almost all categories of physical goods placed on the EU market. The developed frameworks will also provide a platform for establishing horizontal rules. As a first step, it is essential to ensure that a significant number of designed products are more durable, repairable, and recyclable. Additionally, a number of important indicators, such as energy and resource efficiency, need to be measured as they can help reduce the carbon, ecological, and other environmental footprints of products. Companies must compete on a level playing field without engaging in unfair practices that cause environmental damage, while, at the same time, making sure that consumers have access to information about more sustainable choices. Ultimately, people need protection against practices that are harmful to the green transition and limit the availability of longer-lasting products. Furthermore, companies should have access to data ensuring appropriate environmental protection and circularity of their products. By 2030, the new sustainable products framework should generate savings in natural gas in the region of 150 billion m³. Nowadays, this is very important as it can help several EU members reduce their dependence on fossil fuels, including Russian gas.

In the previous sections of this chapter, the authors reviewed the current "take-make-dispose" economic model (European Commission – About Sustainable Products, 2024). By contrast, sustainable products are obtained from circular-economy applications in *cradle-to-cradle* lifecycles (Belz & Peattie, 2012; Braungart et al., 2007; Moldavska & Welo, 2017).

The *cradle-to-cradle* idea (Figure 2.4) is, in fact, a full circle process that begins with the creation of safe products for both human and environmental health (Sherratt, 2013).

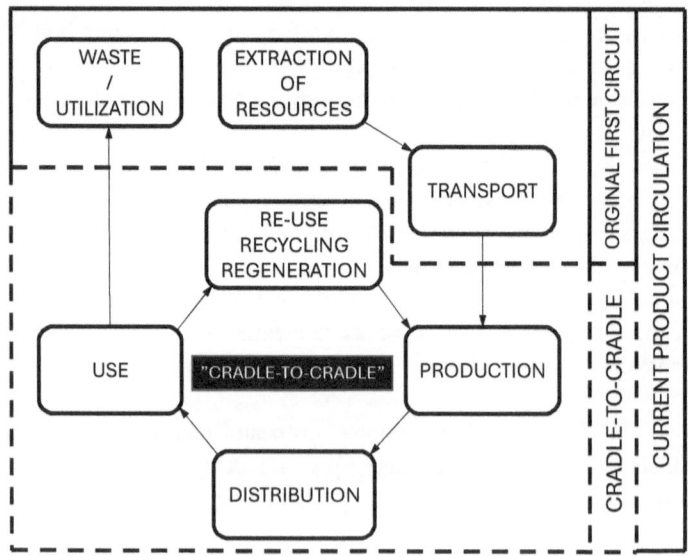

FIGURE 2.4 Product circulation diagram promoting the CRADLE-TO-CRADLE idea of sustainable development.

Source: personal elaboration from (Trojanowski, 2017).

In Figure 2.4, the authors of this chapter presented a diagram showing the current circulation of products and the *cradle-to-cradle* idea (Finnigan et al., 2017; Trojanowski, 2017). The latter is limited to the current four steps: recycling, production, distribution, and use. The first life-cycle step is about extracting the raw materials needed for product making. In the next step, the obtained materials are transported to plants where the finished products are manufactured. The "commodities" are distributed to retailing systems, and then they are purchased by consumers. In the circulation process, products in the cradle-to-cradle model can either be passed for re-use, recycling, or regeneration, or they are sent for utilization.

2.3   Product labelling and consumer decision-making

In contemporary society, decisions made by consumers play a crucial role in shaping the market and influencing production practices (Szcześniak, 2020). The social and environmental factors influencing consumers' choices, along with the consequences of those choices, can have significant implications for the future of our planet (Kiełczewski, 2007). The overall high economic growth trend has led to a significant increase in consumption. The negative consequence of such actions is their harmful impact on the environment. Therefore, as part of efforts to mitigate the adverse effects of excessive consumption, developing a consumer education system is crucial. Such a solution will enhance consumer awareness of the impact of products, enabling the identification of a product's positive or negative environmental effects. One solution enabling this type of identification is appropriate product labelling (Wojnarowska et al., 2021).

Product labels should include information about the sustainable use of raw materials, energy consumption, savings in water, the use of recyclable materials, carbon dioxide emissions, and hazardous chemicals (Paul et al., 2023). The first ecolabel was launched in 1977 by the Blue Angel in Germany. Since 1977, several countries have introduced their own ecolabel programmes, including Canada's Environmental Choice in 1988, Israel's Green Label in 1993, and Hong Kong's Green Label Scheme in 2000 (Korteland, 2007). Hence, such labellings have been in use for about 50 years.

The research conducted to date (Brzostek-Kasprzak et al., 2013) indicates that price is a significant factor shaping demand for specific products. However, over half of surveyed consumers declared that they were willing to pay a higher price for ecologic products. The majority (72%) declared they may be willing to pay a 10% higher price for such products. Some respondents (20%) expressed a possible willingness to pay up to 25% more for ecological products. Unfortunately, market prices for environmentally friendly products are much higher. Due to such economic barriers, some people cannot afford to buy more expensive products. Even though sustainable products might cost more upfront, they usually end up being more economical in the long run. For example, energy-efficient appliances can result in large power bill savings, while long-lasting materials lessen the need for replacements over time. Another study (Mandarić et al., 2022) focused on the specific factors that are taken into account when purchasing goods and products, such as price, value, size, quality, style, convenience of purchase, and materials. On the other hand, for the majority of consumers, environmental factors are not so important. Furthermore, consumer behaviour in the realm of sustainable product consumption often does not align with their actual choices regarding sustainable products. Even when consumers declare their intention to buy environmentally friendly products, this may not at all be reflected in their actual purchasing decisions.

However, several factors shape consumer choices regarding environmentally friendly products. In conclusion, for labelling to yield tangible and measurable effects, three conditions most likely must be met. The first condition is familiarity with the product labelling system and ensuring accurate identification. The second condition is the existence of a suitably wealthy society, thanks to which enough consumers have the financial means needed to purchase environmentally friendly products. The third, and perhaps most crucial aspect, concerns the final decisions made when making purchases. Simply possessing knowledge and the ability to verify product types based on labels, as well as the necessary financial resources, do not guarantee that consumers will opt for environmentally friendly products.

2.4   Conclusions

In summary, it is important to stress the fact that our knowledge about product transformation is broad and complex. The subsections contain specific information about product types. This study provides insights on various issues connected with products and their identification. Overall, this is important in the context of the future development of different types of products.

In the future, the indicators of different product descriptions and their classification should be based on scientific parameters. Only measurable values in quantity form ensure the ultimate clarity and accuracy of information regarding a type of product. This is one of the most important goals that need to be achieved. The current literature on various products and their transformation provides no synthetic descriptions that would enable any classification of product types. Such considerations should be based on evaluation criteria, achievable only through the formulation of a mathematical model. The proposed model should unequivocally facilitate identification of each product type, ranging from conventional to sustainable. Finding such a mathematical formula is necessary and would require the involvement of other scientific disciplines such as Agricultural Sciences, Natural Sciences, Medical & Health Sciences, and Engineering & Technology. Developing appropriate and precise criteria will pave the way for a transparent product certification process. At the same time, laws and regulations should mandate the possibility of labelling every product. This way, consumers will be accurately informed about the type of product they are dealing with and how it affects the environment.

Acknowledgement

This research was funded in whole or in part by National Science Centre, Poland 2022/45/B/HS4/00363. For the purpose of Open Access, the author has applied a CC-BY public copyright licence to any Author Accepted Manuscript (AAM) version arising from this submission

References

Al-Ashaab, A., Golob, M., Attia, U. M., Khan, M., Parsons, J., Andino, A., Perez, A., Guzman, P., Onecha, A., Kesavamoorthy, S., Martinez, G., Shehab, E., Berkes, A., Haque, B., Soril, M., & Sopelana, A. (2013). The transformation of product development process into lean environment using set-based concurrent engineering: A case study from an aerospace industry. *Concurrent Engineering Research and Applications, 21*(4), 268–285. https://doi.org/10.1177/1063293X13495220

Alqalami, T. A., Elkadi, H., & Al-Alwan, H. (2020). The application of BIM tools to explore the dynamic characteristics of smart materials in a contemporary shanashil building design element. *International Journal of Sustainable Development and Planning, 15*(2), 193–199. https://doi.org/10.18280/IJSDP.150209

Azhar, N. N. H., Ang, D. T. C., Abdullah, R., Harikrishna, J. A., & Cheng, A. (2022). Bio-based materials riding the wave of sustainability: Common misconceptions, opportunities, challenges and the way forward. *Sustainability, 14*(9), 5032. https://doi.org/10.3390/SU14095032

Belz, F. M., & Peattie, K. (2012). Customer cost. *Sustainability Marketing: A Global Perspective, 62*, 231–254. https://www.wiley.com/en-us/Sustainability+Marketing%3A+A+Global+Perspective%2C+2nd+Edition-p-9781119966197

Bernardi, B., Falcone, G., Stillitano, T., Benalia, S., Bacenetti, J., & De Luca, A. I. (2021). Harvesting system sustainability in Mediterranean olive cultivation: Other principal cultivar. *Science of the Total Environment, 766*, 142508. https://doi.org/10.1016/J.SCITOTENV.2020.142508

Bhardwaj, A. K., Garg, A., Ram, S., Gajpal, Y., & Zheng, C. (2020). Research trends in green product for environment: A bibliometric perspective. *International Journal of Environmental Research and Public Health, 17*(22), 8469. https://doi.org/10.3390/IJERPH17228469

Biswas, A., & Roy, M. (2015). Green products: An exploratory study on the consumer behaviour in emerging economies of the East. *Journal of Cleaner Production, 87*(1), 463–468. https://doi.org/10.1016/J.JCLEPRO.2014.09.075

Braungart, M., McDonough, W., & Bollinger, A. (2007). Cradle-to-cradle design: Creating healthy emissions – A strategy for eco-effective product and system design. *Journal of Cleaner Production, 15*(13–14), 1337–1348. https://doi.org/10.1016/J.JCLEPRO.2006.08.003

Brzostek-Kasprzak, B., Kwasek, M., Obiedzińska, A., & Obiedziński, M. W. (2013). *Z badań nad rolnictwem społecznie zrównoważonym (21). Żywność ekologiczna – regulacje prawne, system kontroli i certyfikacji.* https://depot.ceon.pl/handle/123456789/5849

Calabrese, A., Costa, R., Ghiron, N. L., Tiburzi, L., & Pedersen, E. R. G. (2021). How sustainable-orientated service innovation strategies are contributing to the sustainable development goals. *Technological Forecasting and Social Change, 169*, 120816. https://doi.org/10.1016/J.TECHFORE.2021.120816

Chia, W. Y., Chew, K. W., Le, C. F., Lam, S. S., Chee, C. S. C., Ooi, M. S. L., & Show, P. L. (2020). Sustainable utilization of biowaste compost for renewable energy and soil amendments. *Environmental Pollution (Barking, Essex : 1987), 267*. https://doi.org/10.1016/J.ENVPOL.2020.115662

Composting At Home | US EPA (2024). https://www.epa.gov/recycle/composting-home

Coronado-Medina, A., Arias-Pérez, J., & Perdomo-Charry, G. (2020). Fostering product innovation through digital transformation and absorptive capacity. *International Journal of Innovation and Technology Management, 17*(6). https://doi.org/10.1142/S0219877020500406

Corrêa, M., Lima, B. V. de M., Martins, V. W. B., Rampasso, I. S., Anholon, R., Quelhas, O. L. G., & Leal Filho, W. (2020). An analysis of the insertion of sustainability elements in undergraduate design courses offered by Brazilian higher education institutions: An exploratory study. *Journal of Cleaner Production, 272*, 122733. https://doi.org/10.1016/J.JCLEPRO.2020.122733

Das, S., Ray, A., & De, S. (2020). Optimum combination of renewable resources to meet local power demand in distributed generation: A case study for a remote place of India. *Energy, 209*, 118473. https://doi.org/10.1016/J.ENERGY.2020.118473

Dastan, S., Ghareyazie, B., & Pishgar, S. H. (2019). Environmental impacts of transgenic Bt rice and non-Bt rice cultivars in northern Iran. *Biocatalysis and Agricultural Biotechnology, 20*, 101160. https://doi.org/10.1016/J.BCAB.2019.101160

Dinu, V., Savoiu, G. G., & Tachiciu, L. (2012). *Eco-Labeling and Consumers' Education Framework. 18th IGWT Symposium, Technology and Innovation for a Sustainable Future: A Commodity Science Perspective,* September 2012. https://papers.ssrn.com/sol3/papers.cfm?abstract_id=2315491

European Bioplastics (2024). *European Bioplastics.* https://www.european-bioplastics.org/

European Commission – About Sustainable Products (2024). *About Sustainable Products.* https://commission.europa.eu/energy-climate-change-environment/standards-tools-and-labels/products-labelling-rules-and-requirements/sustainable-products/about-sustainable-products_en

European Commission – Circular Economy Action Plan (2024). *Circular Economy Action Plan.* https://environment.ec.europa.eu/strategy/circular-economy-action-plan_en

European Commission – Ecodesign for Sustainable Products Regulation (2024). *Ecodesign for Sustainable Products Regulation.* https://commission.europa.eu/energy-climate-change-environment/standards-tools-and-labels/products-labelling-rules-and-requirements/sustainable-products/ecodesign-sustainable-products-regulation_en?prefLang=en

European Commission – Organics at a Glance (2024). *Organics at a Glance.* https://agriculture.ec.europa.eu/farming/organic-farming/organics-glance_en

Ferraris, F., Iacoponi, F., Raggi, A., Baldi, F., Fretigny, M., Mantovani, A., & Cubadda, F. (2021). Essential and toxic elements in sustainable and underutilized seafood species and derived semi-industrial ready-to-eat products. *Food and Chemical Toxicology, 154*, 112331. https://doi.org/10.1016/J.FCT.2021.112331

Finnigan, T., Needham, L., & Abbott, C. (2017). Mycoprotein: A healthy new protein with a low environmental impact. *Sustainable Protein Sources, 305–325.* https://doi.org/10.1016/B978-0-12-802778-3.00019-6

Fook, L. A., & McNeill, L. (2020). Click to buy: The impact of retail credit on over-consumption in the online environment. *Sustainability (Switzerland), 12*(18). https://doi.org/10.3390/SU12187322

Garekwe, L., Ferreira-Schenk, S. J., & Dickason-Koekemoer, Z. (2024). Modelling factors influencing bank customers' readiness for artificial intelligent banking products. *International Journal of Economics and Financial Issues, 14*(1), 73–84. https://doi.org/10.32479/IJEFI.15238

Gerassimidou, S., Martin, O. V., Chapman, S. P., Hahladakis, J. N., & Iacovidou, E. (2021). Development of an integrated sustainability matrix to depict challenges and trade-offs of introducing bio-based plastics in the food packaging value chain. *Journal of Cleaner Production, 286*, 125378. https://doi.org/10.1016/J.JCLEPRO.2020.125378

Herrmann, A., Gassmann, O., & Eisert, U. (2007). An empirical study of the antecedents for radical product innovations and capabilities for transformation. *Journal of Engineering and Technology Management – JET-M, 24*(1–2), 92–120. https://doi.org/10.1016/J.JENGTECMAN.2007.01.006

Hofenk, D., van Birgelen, M., Bloemer, J., & Semeijn, J. (2019). How and when retailers' sustainability efforts translate into positive consumer responses: The interplay between personal and social factors. *Journal of Business Ethics, 156*(2), 473–492. https://doi.org/10.1007/S10551-017-3616-1/FIGURES/2

Jones, S. M., Lesaux, N. K., Gonzalez, K. E., Hanno, E. C., & Guzman, R. (2020). Exploring the role of quality in a population study of early education and care. *Early Childhood Research Quarterly, 53*, 551–570. https://doi.org/10.1016/J.ECRESQ.2020.06.005

Kiełczewski, D. (2007). Struktura pojęcia konsumpcji zrównoważonej. *Ekonomia i Środowisko, 2*(32), 36–50.

Klaus, V. H., & Kiehl, K. (2021). A conceptual framework for urban ecological restoration and rehabilitation. *Basic and Applied Ecology, 52*, 82–94. https://doi.org/10.1016/J.BAAE.2021.02.010

Korteland, M. (2007). Eco-labelling: To be or not to be? *Rep. by CE Delft.* https://cedelft.eu/wp-content/uploads/sites/2/2021/03/07_7479_17.pdf

Lin, D., Hanscom, L., Murthy, A., Galli, A., Evans, M., Neill, E., Mancini, M. S., Martindill, J., Medouar, F. Z., Huang, S., & Wackernagel, M. (2018). Ecological footprint accounting for countries: Updates and results of the national footprint accounts, 2012–2018. *Resources, 7*(3), 58. https://doi.org/10.3390/RESOURCES7030058

Ljungberg, L. Y. (2007). Materials selection and design for development of sustainable products. *Materials & Design, 28*(2), 466–479. https://doi.org/10.1016/J.MATDES.2005.09.006

Lopes, I. G., Braos, L. B., Cruz, M. C. P., & Vidotti, R. M. (2021). Valorization of animal waste from aquaculture through composting: Nutrient recovery and nitrogen mineralization. *Aquaculture, 531*, 735859. https://doi.org/10.1016/J.AQUACULTURE.2020.735859

Machindra Gurme, V. (2022). E-commerce marketplaces selling consumer durable sustainable products in India: A market analysis. *EPRA International Journal of Multidisciplinary Research (IJMR), 8*(5), 1–1. https://doi.org/10.36713/epra2013

Malinowski, L., Kowczyk-Sadowy, M., Obidzinski, S., Krasowska, M., & Dolzynska, M. (2019). Żywność z upraw ekologicznych i konwencjonalnych oraz sposoby jej znakowania. *Technika Rolnicza Ogrodnicza Leśna, 4*, 17–20.

Mandarić, D., Hunjet, A., & Vuković, D. (2022). The impact of fashion brand sustainability on consumer purchasing decisions. *Journal of Risk and Financial Management, 15*(4), 176. https://doi.org/10.3390/JRFM15040176

Markeset, T., & Kumar, U. (2005). Product support strategy: Conventional versus functional products. *Journal of Quality in Maintenance Engineering, 11*(1), 53–67. https://doi.org/10.1108/13552510510589370

Meier, M. S., Stoessel, F., Jungbluth, N., Juraske, R., Schader, C., & Stolze, M. (2015). Environmental impacts of organic and conventional agricultural products – Are the differences captured by life cycle assessment? *Journal of Environmental Management, 149*, 193–208. https://doi.org/10.1016/J.JENVMAN.2014.10.006

Moldavska, A., & Welo, T. (2017). The concept of sustainable manufacturing and its definitions: A content-analysis based literature review. *Journal of Cleaner Production, 166*, 744–755. https://doi.org/10.1016/J.JCLEPRO.2017.08.006

Nuryakin, N. & Maryati, T. (2020). Green product competitiveness and green product success. Why and how does mediating affect green innovation performance? *Entrepreneurship and Sustainability Issues, 7*(4), 3061–3077. https://doi.org/10.9770/JESI.2020.7.4(33)

Okogwu, C., Odochi Agho, M., Abimbola Adeyinka, M., Odulaja, B. A., Louis Eyo-Udo, N., Daraojimba, C., & Alex Banso, A. (2023). Exploring the integration of sustainable materials in supply chain management for environmental impact. *Engineering Science & Technology Journal, 4*(3), 49–65. https://doi.org/10.51594/ESTJ.V4I3.546

Ontario. (2024). *Introduction to Bioproducts.* https://www.ontario.ca/page/introduction-bioproducts#Ref1

Ottman, J. A., Stafford, E. R., & Hartman, C. L. (2006). Avoiding green marketing myopia: Ways to improve consumer appeal for environmentally preferable products. *Environment, 48*(5), 22–36. https://doi.org/10.3200/ENVT.48.5.22-36/ASSET//CMS/ASSET/9536EF8D-8E3A-4B6B-AF9C-B3C55244F0D3/ENVT.48.5.22-36.FP.PNG

Palhares, J. C. P., Morelli, M., & Novelli, T. I. (2021a). Water footprint of a tropical beef cattle production system: The impact of individual-animal and feed management. *Advances in Water Resources, 149*, 103853. https://doi.org/10.1016/J.ADVWATRES.2021.103853

Palhares, J. C. P., Morelli, M., & Novelli, T. I. (2021b). Water footprint of a tropical beef cattle production system: The impact of individual-animal and feed management. *Advances in Water Resources, 149*, 103853. https://doi.org/10.1016/J.ADVWATRES.2021.103853

Paul, D., Malik, S., Mishra, D. K., & Teotia, A. (2023). A study on effectiveness of ecolabels in the fast-moving consumer goods sector. *IOP Conference Series: Earth and Environmental Science, 1161*(1), 012004. https://doi.org/10.1088/1755-1315/1161/1/012004

Qiu, J., & Cardinale, B. J. (2020). Scaling up biodiversity–ecosystem function relationships across space and over time. *Ecology, 101*(11), e03166. https://doi.org/10.1002/ECY.3166

Rahman, M. M., Aravindakshan, S., Hoque, M. A., Rahman, M. A., Gulandaz, M. A., Rahman, J., & Islam, M. T. (2021). Conservation tillage (CT) for climate-smart sustainable intensification: Assessing the impact of CT on soil organic carbon accumulation, greenhouse gas emission and water footprint of wheat cultivation in Bangladesh. *Environmental and Sustainability Indicators, 10*, 100106. https://doi.org/10.1016/J.INDIC.2021.100106

Roome, N., & Anastasiou, I. (2002). Sustainable production: Challenges and objectives for EU Research Policy. *Reflets et Perspectives de La Vie Economique, 41*(1), 35–49. https://doi.org/10.3917/RPVE.411.0035

Rosa, F. S. da, Lunkes, R. J., Spigarelli, F., & Compagnucci, L. (2021). Environmental innovation and the food, energy and water nexus in the food service industry. *Resources, Conservation and Recycling, 166*, 105350. https://doi.org/10.1016/J.RESCONREC.2020.105350

Sdrolia, E., Zarotiadis, G., Sdrolia, E., & Zarotiadis, G. (2019). A comprehensive review for green product term: From definition to evaluation. *Journal of Economic Surveys, 33*(1), 150–178. https://doi.org/10.1111/JOES.12268

Sengstschmid, H., Sprong, N., Schmid, O., Stockebrand, N., Stolz, H., Spiller, A., & Fitzsimons, D. (2011). *For European Commission Oakdene Hollins Provides Clients with These Services.* www.oakdenehollins.co.uk

Sharma, H. B., Vanapalli, K. R., Samal, B., Cheela, V. R. S., Dubey, B. K., & Bhattacharya, J. (2021). Circular economy approach in solid waste management system to achieve UN-SDGs: Solutions for post-COVID recovery. *Science of the Total Environment, 800*, 149605. https://doi.org/10.1016/J.SCITOTENV.2021.149605

Sherratt, A. (2013). Cradle to cradle. *Encyclopedia of Corporate Social Responsibility, 630–638.* https://doi.org/10.1007/978-3-642-28036-8_165

Smith, L., Ibn-Mohammed, T., Astudillo, D., Brown, S., Reaney, I. M., & Koh, S. C. L. (2021). The role of cycle life on the environmental impact of Li6.4La3Zr1.4Ta0.6O12 based solid-state batteries. *Advanced Sustainable Systems*, *5*(2), 2000241. https://doi.org/10.1002/ADSU.202000241

Stolz, H., Stolze, M., Janssen, M., & Hamm, U. (2011). Preferences and determinants for organic, conventional and conventional-plus products – The case of occasional organic consumers. *Food Quality and Preference*, *22*(8), 772–779. https://doi.org/10.1016/J.FOODQUAL.2011.06.011

Szcześniak, M. (2020). Proces decyzyjny konsumenta na rynku. *Zeszyty Naukowe PUNO*, *8*(1), 379–390.

Tezer, A., & Bodur, H. O. (2020). The greenconsumption effect: How using green products improves consumption experience. *Journal of Consumer Research*, *47*(1), 25–39. https://doi.org/10.1093/JCR/UCZ045

Thangaraj, P., Gordon, B., Zilberma, D. N., Hochman, G., Zilberman, D., & Wang, D. (2018). *What Role Do Bioproducts Play? Biomass to Bioproducts: Economic and Environmental Viability*. https://doi.org/10.3923/jm.2015.181.192

Țigan, E., Brînzan, O., Obrad, C., Lungu, M., Mateoc-Sîrb, N., Milin, I. A., & Gavrilaș, S. (2021). The consumption of organic, traditional, and/or European eco-label products: Elements of local production and sustainability. *Sustainability*, *13*(17), 9944. https://doi.org/10.3390/SU13179944

Trojanowski, T. (2017). Projektowanie zrównoważonych produktów. *Zeszyty Naukowe. Organizacja i Zarządzanie / Politechnika Śląska, z. 100*, 513–522.

Wan, N. F., Su, H., Cavalieri, A., Brack, B., Wang, J. Y., Weiner, J., Fan, N. N., Ji, X. Y., & Jiang, J. X. (2020). Multispecies co-culture promotes ecological intensification of vegetable production. *Journal of Cleaner Production*, *257*, 120851. https://doi.org/10.1016/J.JCLEPRO.2020.120851

Wojnarowska, M., Ćwiklicki, M., & Ingrao, C. (2022). Sustainable products in the circular economy: Impact on business and society. *Sustainable Products in the Circular Economy: Impact on Business and Society*, 1–280. https://doi.org/10.4324/9781003179788

Wojnarowska, M., Sołtysik, M., & Prusak, A. (2021). Impact of eco-labelling on the implementation of sustainable production and consumption Keywords: Sustainable production Sustainable consumption Eco-labelling. *Environmental Impact Assessment Review*, *86*, 106505. https://doi.org/10.1016/j.eiar.2020.106505

Xu, Y., & Szmerekovsky, J. (2017). System dynamic modeling of energy savings in the US food industry. *Journal of Cleaner Production*, *165*, 13–26. https://doi.org/10.1016/J.JCLEPRO.2017.07.093

Yadav, D., Kumari, R., Kumar, N., & Sarkar, B. (2021). Reduction of waste and carbon emission through the selection of items with cross-price elasticity of demand to form a sustainable supply chain with preservation technology. *Journal of Cleaner Production*, *297*, 126298. https://doi.org/10.1016/J.JCLEPRO.2021.126298

Yusoff, Y. M., Omar, M. K., Kamarul Zaman, M. D., & Samad, S. (2019). Do all elements of green intellectual capital contribute toward business sustainability? Evidence from the Malaysian context using the Partial Least Squares method. *Journal of Cleaner Production*, *234*, 626–637. https://doi.org/10.1016/J.JCLEPRO.2019.06.153

Zhang, Q., Yang, Y., Shang, N., Xiao, Y., Xiao, Y., Liu, Y., Jiang, X., Sanganyado, E., Liu, S., & Xia, X. (2023). Identification and coexposure of neonicotinoid insecticides and their transformation products in retail cowpea (*Vigna unguiculata*). *Environmental Science and Technology*, *57*(48), 20182–20193. https://doi.org/10.1021/ACS.EST.3C05269

3

DESIGNING AND DEVELOPING SUSTAINABLE PRODUCTS

Marcin Paprocki

3.1   Introduction

In the face of increasing environmental degradation and climate change-related issues, the need for a shift away from conventional products to ecological and sustainable alternatives is gaining in importance. This transformation is not only impacting the products themselves but also necessitating modifications in stakeholders' approach towards their consumption, design, production, and development, as well as changes in the regulatory framework. As society becomes increasingly environmentally conscious, there is a growing demand for ecological and bio products. Consumers are showing a greater tendency to purchase products that have a less negative impact on the environment.

Ecological products can be defined as products made from recyclable materials and produced using water and energy-saving methods to limit waste, packaging, and the time required for the disposal of toxic materials (Nimse et al., 2007). In turn, bio-products are materials, chemicals, and energy derived from renewable biological material (Singh et al., 2003). On the other hand, stakeholders are becoming increasingly conscious of the need to care about social growth and justice. In this context, products should have a positive impact on social aspects throughout their entire life cycle.

As a result of both these trends, there is a growing realization among entrepreneurs that they need to adapt the products they offer to consumers in line with changing market preferences and legal regulations. According to Realyvásquez-Vargas et al. (2023), companies must streamline their production processes to increase the scope of sustainable development (from an economic, environmental, social, and technological perspective, as well as in terms of efficiency, energy consumption, performance management, production, and quality).

DOI: 10.4324/9781032710693-4

3.2   Design and development of sustainable products

Given the expectations outlined above, many trends in product design and development can be observed. The products that stand out in this light are innovative, ecological, sustainable, cheap, high-quality, fashionable, tailored to the customer's needs and expectations, created within the circular economy (CE), take into account the concept of corporate social responsibility (CSR), and are created according to the principles of Industry 4.0.

Sustainable design is essentially the process of developing a product that successfully performs its intended functions, generates profits for the company, is socially acceptable, and consumes a minimum amount of energy and materials without producing hazardous waste (Chiu & Chu, 2012).

The task of managing the development of sustainable products should encompass all stages of their life cycle as well as take into account the needs and interests of partners, suppliers, and transportation companies, as well as other participants involved in the development and product life cycle. Those stages during which the greatest environmental impact occurs in the product life cycle are the following (Paprocki, 2017b): raw material extraction; the production of materials, energy, components, and parts; the manufacturing of products, distribution, and transportation of products; their usage and servicing; and their recycling and disposal.

On the other hand, the stage in the product life cycle during which the characteristics of a product are shaped most significantly, and upon which its quality most depends, is the production preparation stage (pre-production stage). During this stage, the ecological aspects of the product are also designed/shaped, and other aspects of sustainable product development are taken into account.

Roughly 80% of all forms of environmental impact associated with the manufacturing process are identified during the product design phase (Schischke et al., 2005; United Nations Environment Programme & International Resource Panel, 2018). The designed product must primarily be safe and not pose a threat to the health and life of users. It should satisfy their expectations regarding utility and functionality. Subsequently, the quality of products and manufacturing processes, as well as their reliability and ecological sustainability, are crucial. During the design process, efforts should also be made to keep production costs down to ensure profitability.

A key aspect of product development, especially in the context of meeting the requirements and expectations of various stakeholders, is its balancing. When considering customer expectations, it is worth defining the directions and extent of product balancing. The latter provides the basis for effective management of sustainable product development, and the best option can be chosen from various sustainable development alternatives for the product. The processes shaping the design and development of a sustainable product are illustrated in Figure 3.1.

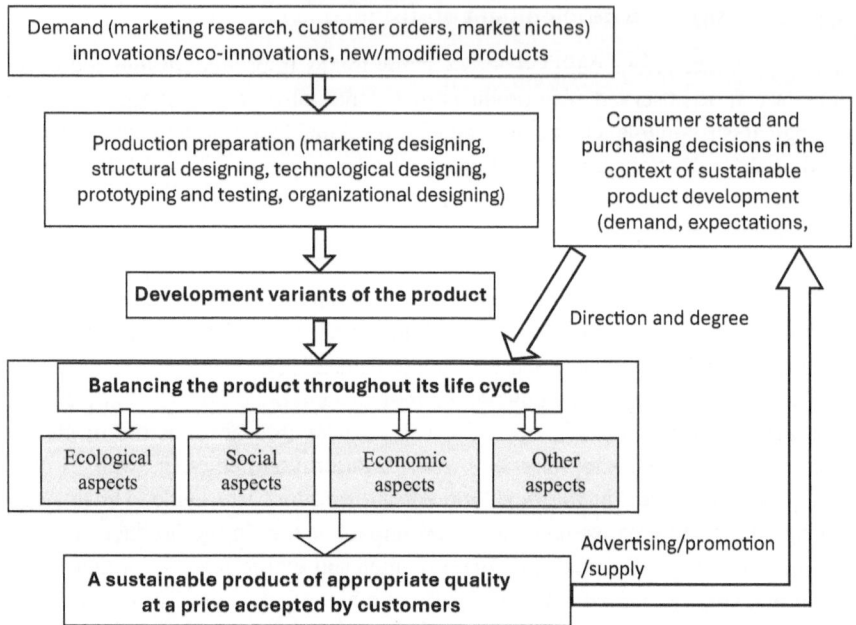

FIGURE 3.1 Designing and developing sustainable products.

Source: Author's own diagram.

3.3   Fast prototyping and manufacturing techniques: RP/RT and RM

Rapid additive manufacturing techniques, commonly known as three-dimensional (3D) printing, include rapid prototyping (RP), rapid manufacturing (RM), and rapid tooling (RT). These techniques, utilizing methods such as 3D printing, hybrid printing, and reverse engineering, support the design and manufacturing of products, especially in unit and small batch production. The use of 3D printing in the prototyping (RP) and manufacturing (RM) stages facilitates the design, manufacturing, and development of sustainable products (Embia et al., 2023; Kerry Taylor-Smith, 2021), as well as the achievement of sustainable development goals (SDGs) – particularly the following goals: SDG 1 (no poverty), SDG 3 (good health and wellbeing), SDG 4 (quality education), SDG 9 (industry, innovation, and infrastructure), SDG 12 (responsible consumption and production), and SDG 14 (life below water) (Muth et al., 2023). According to a study conducted at Michigan Technological University, printing an object in 3D form consumes 4164% less energy than producing it overseas and shipping it to the United States. This was due to the reduced volume of required materials and a reduced demand for shipping (Kerry Taylor-Smith, 2021).

3.4 Tools and methods supporting the design and development of sustainable products

3.4.1 *Environmental management tools*

When it comes to environmental management tools, the life cycle assessment (LCA) technique is of particular importance. LCA provides effective support for eco-design management and product development by taking into account ecological aspects. LCA can also be successfully used to compare both existing products and those still in the design stage in terms of their environmental impact throughout their life cycle. Another of its functions is to assess variants of designed products so as to choose the best option (or options).

The ecological assessment of a product's life cycle is divided into four phases: (I) goal and scope definition, (II) life cycle inventory analysis, (III) life cycle impact assessment, and (IV) interpretation (ISO, 2006). LCA has been implemented in various software systems. The most popular LCA software, tailored to European specifications, includes programs such as SimaPro, GaBi, and Umberto. When managing sustainable product development, the environmental, social, and financial aspects of the process need to be taken into account. Environmental aspects are assessed using LCA, while economic aspects are evaluated using environmental life cycle costing (E-LCC), and social aspects by means of social life cycle assessment (S-LCA).

Environmental management tools and techniques, such as LCA and LCC, are described in more detail in Chapter 4.

3.4.2 *Design for X methods*

Methods of design for X (DFX) are utilized to enhance specific aspects of the developed product. X is typically replaced with the optimization target, and these methodologies are employed to support the product development process (Formentini et al., 2022). Researchers (Chiu & Chu, 2012) have focused on the perspectives of DFX design related to sustainable product development, such as design for manufacturing (DFM), design for environment (DFE), design for disassembly (DFD), and design for recycling (DFR).

Currently, an increasing number of companies are investing in the design of environmentally friendly products with a view to boosting their brand competitiveness and developing ecological products (Wang et al., 2023). In this context, it is worth noting certain design tools from the perspective of their impact on the natural environment, such as DFE.

DFE is an environmental management feature that addresses environmental issues during the product or process design stage (Puglieri et al., 2020; Zhang et al., 2019). DFE encompasses various tools, methods, and principles whose purpose is to assist designers in reducing the environmental impact of the products they design (Telenko et al., 2016). The authors of the study proved the hypothesis that DFE combined with quality management innovation can play a strategic role in supporting the effective implementation of environmentally friendly production practices in the areas of design, production, delivery, use, recycling/remanufacturing, and disposal of products with the aim of achieving better results in the field of environmental protection (Jackson et al., 2016). The various components of DFE may include, for example, methodologies oriented towards ease of disassembly (DFD), recycling (DFR), remanufacturing (DFRem), and product longevity (DFL).

DFD is a product design method utilized by designers to enable and streamline the disassembly process.

Disassembly strategies have become crucial, particularly for end-of-life products, where separating parts made from different materials is paramount. Therefore, incorporating and planning correct disassembly procedures at an early stage of the design process can lead to improved sustainable product development (Paprocki, 2017). By employing DFD to support sustainable design, the amount of waste and scrap products remaining after their withdrawal is reduced, enabling the reuse, regeneration, and recycling of products (Abuzied et al., 2020; Mule, 2012).

DFR is a design method that promotes the recyclability of products at the end of their life by facilitating the separation of different materials and ensuring comprehensive, efficient material utilization.

A complement to, and an extension of designing for recycling is the design from recycling (DFromR) strategy, according to which secondary raw materials serve as the starting point for developing a new product (Ragaert et al., 2020). In this context, from the perspective of a product designer, two approaches can be applied (Leal et al., 2020):

- Design for end of life is the best known and most commonly applied approach. Its aim is to improve a product so that it can be recovered in the best possible way when it becomes waste, as well as to facilitate the elimination of product residues that cannot be recovered.
- Design from end of life involves integrating artefacts from the processing chain after their end of use into new products (e.g., using recycled materials instead of primary ones, reusing modules or parts obtained during disassembly).

From the perspective of sustainable environmental development, extending the lifespan of a product through regeneration is a promising approach to reducing both pollution, especially CO_2 emissions, as well as our dependence on primary materials of geological origin (Boorsma et al., 2021). One tool that helps extend lifespan in product design is DFRem.

The goal of DFRem is to design a product in a way that facilitates its regeneration by, for example, simplifying the process of its disassembly, cleaning, reprocessing, and reassembly (Lindkvist Haziri & Sundin, 2020; Subramoniam et al., 2010).

Extending the lifespan of products, such as clothing, is key to reducing their overall environmental impact (Laitala & Klepp, 2021). One tool that supports this process at the design stage is DFL – designing for the product's lifespan.

The premise of DFL involves undertaking various efforts aimed at extending a product's lifespan (Klepp et al., 2020). The goal of DFL is to design products with an optimal lifespan, where optimal in this sense means taking into account the perspectives of users, businesses, and resource efficiency during a product's life cycle design (Nurcahyanie & Rohmadiani, 2023). DFL gears design activities towards product sustainability in terms of lifespan.

Another important tool supporting the design and sustainable development of products is the design for sustainability (DFS) method. This method involves incorporating sustainable development goals into the design process. This threefold approach involves balancing traditional economic goals with social and environmental concerns (Browne, 2002; Rosen & Kishawy, 2012). DFS can, and should, play an important role in the transition towards the sustainable production and consumption of production as key elements of quality of life (Spangenberg et al., 2010).

3.4.3   Conventional design tools

Other tools capable of supporting pro-ecological product design include such conventional design methods as (Paprocki, 2017b):

- Quality function deployment (QFD) – developing product quality functions,
- Failure mode and effects analysis (FMEA) – enabling analysis of different types of errors and their impact on the product design and manufacturing processes,
- Design for manufacture and assembly (DFMA) – design oriented towards production and assembly,
- Design of experiments (DOE) – experimental design.

QFD is a method for developing product quality functions, also known as customer driven engineering and matrix product planning. It promotes quality design in new products as well as improvements in existing ones, with a particular focus on customer needs and requirements. The QFD method enables the transformation of market information expressed in the language of consumers, that is, the voice of the customer, into technical language used within a company by designers, engineers, and technologists (Bevilacqua et al., 2006; Karsak, 2004). QFD has been extended to environmental QFD in order to support eco-design. The authors of the publication (Puglieri et al., 2020) identified 29 eco-design methods based on QFD. These methods include both simple house of quality applications and more complex methods involving LCA and LCC analysis.

The FMEA method can be used for eco-design purposes (Puglieri & Ometto, 2011). It makes it possible to predict the probability of failure, assess its effects, identify its causes, and develop preventive, mitigating, and corrective measures to minimize the likelihood of errors in the design and manufacturing processes. To determine which existing or potential defects pose the greatest risk and, based on this, establish the order of preventive and corrective actions, FMEA calculates the risk priority number (RPN) as the product (Aguirre et al., 2021; Braband, 2004) of the following:

- The severity of the defect for the product user or for the result of the production process (number S),
- The probability of the defect occurring (number P),
- The ease with which the defect can be detected (number D).

According to the formula:

$$RPN = S \times P \times D$$

The numbers S, P, and D should be assigned integer values from 1 to 10, so RPN can take values from 1 to 1,000.

The DFMA methodology was developed by G. Boothdrayd and P. Dewhurst based on systems supporting the evaluation of solutions connected with the following:

- Design for assembly,
- DFM.

This methodology has been developed as a software application and helps reduce product development costs by analysing the correctness of manufacturing and assembly processes during the design phase. DFMA is based on systems that assess solutions in terms of assembly and production methods.

DFMA provides, among other things, the following benefit: it reduces product assembly costs as well as the total cost of producing parts by simplifying the structure and selecting the best manufacturing technologies (Paprocki, 2017b). DFMA is an effective tool for simplifying design, production, and assembly, saving time, reducing manufacturing and assembly costs, improving quality, and reducing the environmental impact of products (Leminen et al., 2013; Selvaraj et al., 2009; Yuan et al., 2018).

One team of researchers (Suresh et al., 2016) redesigned an alternator pulley using DFMA and DFE methods. After taken into account dimensional changes in the proposed pulley, it was possible to reduce the environmental impact of the proposed pulley design. In this project, DFE, DFMA, and sustainable development were integrated using the computer aided design (CAD) and ANSYS tools (a calculation package based on the finite element method as part of a computer aided engineering [CAE] approach) to design a sustainable product. It was found that the environmental impact of the redesigned product was minimal.

Similar tools and methods (CAD, CAE, DFMA, and DFE) supporting the design of sustainable products were employed to redesign the connecting rod cap – a component of a four-stroke gasoline engine. The results of the study concluded that implementation of DFE and DFMA may initiate new solutions in the field of sustainable product design with minimal environmental impact (Suresh & Natarajan, 2015).

DOE is a statistical tool applied in various fields of science and industry to support the design, development, and optimization of systems, products, and processes (Durakovic, 2017; Jankovic et al., 2021). DOE techniques can be applied within the sustainable product development design process. In research aimed at shortening the project implementation time and reducing the number of headset defects, the following tools were employed: FMEA, DOE, and the analysis of variance programme (Realyvásquez-Vargas et al., 2023).

The application of tools and methods for assessing the eco-design and life cycle of products supports their design and sustainable development. This translates into an enhanced ability to choose the best sustainable development variant of a product (Figure 3.2).

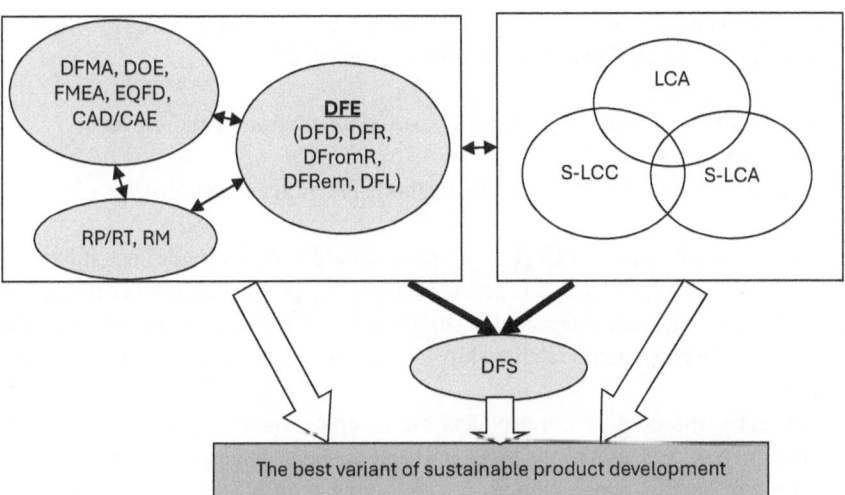

FIGURE 3.2 Tools and methods supporting the design and development of sustainable products.

Source: Author's own diagram.

3.5   Concepts supporting the design and development of a sustainable product

One important concept supporting sustainable product development is the CE model (Diaz et al., 2021). CE promotes circular flows with the aim of reducing environmental impact and maximizing resource efficiency as a sustainable development strategy. It aims to achieve economic prosperity while maintaining environmental quality and social justice so as to create a sustainable world for future generations (Kirchherr et al., 2017).

The goal of the CE system is to preserve added value in products for as long as possible and completely eliminate waste, thereby conserving resources (Zarębska, 2017). The CE involves closing the product life cycle, such that products do not end up in landfills or, worse, are discarded in unauthorized locations (e.g., forests, rivers, seas, oceans) after their use. Instead, efforts are made to extensively reuse, refurbish, recover, and recycle the products.

When the CE system is considered in the context of products, two distinct cycles can be identified (Foundation, 2014). On the one hand, there is the technical cycle, in which technical products are made from extracted resources and produced technical materials. On the other, there is the biological cycle, in which products are made from biological resources obtained from agriculture and from gathering biological materials.

An important method for achieving sustainable development is the concept of CSR. CSR can mitigate corporate harm by encouraging socially responsible and environmentally friendly actions (Shayan et al., 2022; Yadav et al., 2020). According to the principles of CSR, the main areas of corporate responsibility, in order, include the following (Niedek & Hoffmann-Niedek, 2014):

- The product – its quality parameters, including environmental aspects.
- The employees.
- The social environment, including business and other stakeholders.

Following the principles of CSR, a company should aim to provide customers with sustainable goods (products) and services. These are goods and services that, while meeting high standards in terms of customer needs, do not negatively impact social relations or the consumer's relationship with the natural environment (Kiełczewski, 2011).

Currently, more and more products are complex entities, manufactured using numerous resources, materials, parts, and components. Their production is carried out through lengthy production chains involving multiple suppliers, contractors, subcontractors, collaborative entities, and mining companies.

In this context, the concept of the extended enterprise is very helpful for identifying developmental variants of complex products and managing their development, including in sustainable form (Paprocki, 2017a).

In extended enterprises, manufacturing companies closely collaborate with all entities involved to maximize business benefits (Childe, 1998). Therefore, the purpose of applying the above concept is to make the operation of the entire "extended enterprise" as efficient as possible, while optimizing the value added to the product as expected both by the buyer and by the individual entities participating in this project.

Another factor facilitating the development of sustainable products is innovation, especially eco-innovation (Dangelico & Pujari, 2010; Shu et al., 2016). This applies to eco-innovation in the following areas: products and services (product innovations), manufacturing technologies (technological innovations), as well as processes and organizational systems (organizational innovations).

Business process modelling notation (BPMN) is used to model business processes and prepare business diagrams (flowcharts). It is used directly by stakeholders who design, manage, and execute business processes; at the same time, it is precise enough to allow BPMN diagrams to be translated into software process components (Object Management Group, 2013).

The basic components of the notation are as follows (Object Management Group, 2013; Silver, 2011):

• Process flow: events, activities, gateways, artefacts,
• Connections are used to link objects: sequence flow, message flow, associations,
• Process locations: pools, lanes.

The purpose of this notation is to model not only business processes but also engineering processes. It has been implemented in software form, such as iGrafx Process and Bizagi. It is suitable for modelling the different stages of product life cycles. For example, it may serve as a tool for supporting product development design. The author (Paprocki, 2016) used the BPMN notation to model concurrent construction-technological designs of a product. It allows the user to model, simulate, and conduct experiments on various product development variants. Thanks to this, it is possible to design and select the best sustainable development variant for a product, among other things.

3.6   Conclusions

This chapter looked at various aspects of the design and development of a sustainable product, for example, techniques, tools, methods, and notations, as well as approaches that can facilitate the task of designing sustainable products, including manufacturing processes. It described, among other things, a number of fast and innovative techniques: RP, RM, and RT. This chapter also discussed the DFS method alongside the DFE category of environmental design methods as well as design in terms of specific features or properties of sustainable products. It also presented a number of conventional product design methods. The BPMN notation – used for modelling and improving the stages of the product life cycle – was discussed, as was the concept of the extended enterprise, from the perspective of the possibility of modelling, selecting and implementing the best variants of sustainable product development. Concepts supporting the designing and developing sustainable products were also addressed, including the concept of CSR – an approach to design from the social side and the concept of a CE – an approach to design from the environmental side. To summarize, the following (implementation) conclusions can be drawn:

- The use of rapid and innovative methods such as RP and RM support the design, manufacturing, and development of sustainable products.
- One of the most important DFX methods supporting the design and sustainable development of products is DFS.
- DFE methods such as DFD, DFR, DFromR, DFRem, and DFL, especially when combined with LCA, E-LCC, and S-LCA analyses, make it possible to design and select the best variant of sustainable product development.
- Conventional design methods, such as DFMA, DOE, FMEA, QFD, and CAD/CAE, can support product design and sustainable development.
- The BPMN notation, used to model and improve different stages in the product life cycle, can be used to model the stages of sustainable product development.

Acknowledgement

"The publication/article presents the result of the Project no 060/ZJE/2023/POT financed from the subsidy granted to the Krakow University of Economics".

References

Abuzied, H., Senbel, H., Awad, M., & Abbas, A. (2020). A review of advances in design for disassembly with active disassembly applications. *Engineering Science and Technology, an International Journal*, *23*(3). https://doi.org/10.1016/j.jestch.2019.07.003

Aguirre, P. A. G., Pérez-Domínguez, L., Luviano-Cruz, D., Noriega, J. J. S., Gómez, E. M., & Callejas-Cuervo, M. (2021). PFDA-FMEA, an integrated method improving FMEA assessment in product design. *Applied Sciences (Switzerland)*, *11*(4). https://doi.org/10.3390/app11041406

Bevilacqua, M., Ciarapica, F. E., & Giacchetta, G. (2006). A fuzzy-QFD approach to supplier selection. *Journal of Purchasing and Supply Management*, *12*(1). https://doi.org/10.1016/j.pursup.2006.02.001

Boorsma, N., Balkenende, R., Bakker, C., Tsui, T., & Peck, D. (2021). Incorporating design for remanufacturing in the early design stage: A design management perspective. *Journal of Remanufacturing*, *11*(1). https://doi.org/10.1007/s13243-020-00090-y

Braband, J. (2004). Definition and analysis of a new risk priority number concept. *Probabilistic Safety Assessment and Management*. https://doi.org/10.1007/978-0-85729-410-4_322

Browne, J. (2002). Design for the triple top line: New tools for sustainable commerce. *Corporate Environmental Strategy, 9*(3). https://doi.org/10.1016/S1066-7938(02)00069-6

Childe, S. J. (1998). The extended enterprise—A concept of co-operation. *Production Planning and Control, 9*(4). https://doi.org/10.1080/095372898234046

Chiu, M. C., & Chu, C. H. (2012). Review of sustainable product design from life cycle perspectives. *International Journal of Precision Engineering and Manufacturing, 13*(7). https://doi.org/10.1007/s12541-012-0169-1

Dangelico, R. M., & Pujari, D. (2010). Mainstreaming green product innovation: Why and how companies integrate environmental sustainability. *Journal of Business Ethics, 95*(3). https://doi.org/10.1007/s10551-010-0434-0

Diaz, A., Schöggl, J. P., Reyes, T., & Baumgartner, R. J. (2021). Sustainable product development in a circular economy: Implications for products, actors, decision-making support and life cycle information management. *Sustainable Production and Consumption, 26.* https://doi.org/10.1016/j.spc.2020.12.044

Durakovic, B. (2017). Design of experiments application, concepts, examples: State of the art. *Periodicals of Engineering and Natural Sciences, 5*(3), 421–439. https://doi.org/10.21533/pen.v5i3.145; https://www.mendeley.com/catalogue/5084cbff-09ad-3532-9915-7caa8e6edb8e/

Embia, G., Moharana, B. R., Mohamed, A., Muduli, K., & Muhammad, N. B. (2023). *3D Printing pathways for sustainable manufacturing.* https://doi.org/10.1007/978-3-031-20443-2_12

Ellen MacArthur Foundation (2013). *Towards the circular economy Vol. 1: Economic and business rationale for an accelerated transition.* https://ellenmacarthurfoundation.org/towards-the-circular-economy-vol-1-an-economic-and-business-rationale-for-an (Accessed on 7 October 2024).

Formentini, G., Boix Rodríguez, N., & Favi, C. (2022). Design for manufacturing and assembly methods in the product development process of mechanical products: A systematic literature review. *International Journal of Advanced Manufacturing Technology, 120*(7–8). https://doi.org/10.1007/s00170-022-08837-6

ISO. (2006). *ISO 14040 international standard. Environmental management — Life cycle assessment—Principles and framework.* International Organization for Standardization (ISO), Geneva.

Jackson, S. A., Gopalakrishna-Remani, V., Mishra, R., & Napier, R. (2016). Examining the impact of design for environment and the mediating effect of quality management innovation on firm performance. *International Journal of Production Economics, 173.* https://doi.org/10.1016/j.ijpe.2015.12.009

Jankovic, A., Chaudhary, G., & Goia, F. (2021). Designing the design of experiments (DOE) – An investigation on the influence of different factorial designs on the characterization of complex systems. *Energy and Buildings, 250.* https://doi.org/10.1016/j.enbuild.2021.111298

Karsak, E. E. (2004). Fuzzy multiple objective programming framework to prioritize design requirements in quality function deployment. *Computers and Industrial Engineering, 47*(2–3). https://doi.org/10.1016/j.cie.2004.06.001

Kiełczewski, D. (2011). Związki idei zrównoważonego rozwoju z ideą społecznej odpowiedzialności biznesu. *Optimum. Studia Ekonomiczne, 6*(54), 21–31.

Kirchherr, J., Reike, D., & Hekkert, M. (2017). Conceptualizing the circular economy: An analysis of 114 definitions. *Resources, Conservation and Recycling, 127.* https://doi.org/10.1016/j.resconrec.2017.09.005

Klepp, I. G., Laitala, K., & Wiedemann, S. (2020). Clothing lifespans: What should be measured and how. *Sustainability (Switzerland)*, *12*(15). https://doi.org/10.3390/su12156219

Laitala, K., & Klepp, I. G. (2021). Clothing longevity: The relationship between the number of users, how long and how many times garments are used. *4th PLATE 2021 Virtual Conference*. Limerick, Ireland, 26–28 May 2021.

Leal, J. M., Pompidou, S., Charbuillet, C., & Perry, N. (2020). Design for and from recycling: A circular ecodesign approach to improve the circular economy. *Sustainability (Switzerland)*, *12*(23). https://doi.org/10.3390/su12239861

Leminen, V., Eskelinen, H., Matthews, S., & Varis, J. (2013). Development and utilization of a DFMA-evaluation matrix for comparing the level of modularity and standardization in clamping systems. *Mechanika*, *19*(6). https://doi.org/10.5755/j01.mech.19.6.5999

Lindkvist Haziri, L., & Sundin, E. (2020). Supporting design for remanufacturing – A framework for implementing information feedback from remanufacturing to product design. *Journal of Remanufacturing*, *10*(1). https://doi.org/10.1007/s13243-019-00074-7

Mule, J. Y. (2012). Design for disassembly approaches on product development. *International Journal of Scientific & Engineering Research*, *3*(6), 996–1000.

Muth, J., Klunker, A., & Völlmecke, C. (2023). Putting 3D printing to good use—Additive manufacturing and the sustainable development goals. *Frontiers in Sustainability*, *4*. https://doi.org/10.3389/frsus.2023.1196228

Niedek, M., & Hoffmann-Niedek, A. (2014). Produkcja ekologiczna zrównoważona w świetle odpowiedzialności biznesu. *Optimum. Studia Ekonomiczne*, *4*(70), 46–60.

Nimse, P., Vijayan, A., Kumar, A., & Varadarajan, C. (2007). A review of green product databases. *Environmental Progress*, *26*(2). https://doi.org/10.1002/ep.10210

Nurcahyanie, Y. D., & Rohmadiani, L. D. (2023). Design for longevity and design for X: Concepts, applications, and perspectives. *Tibuana*, *6*(1). https://doi.org/10.36456/tibuana.6.1.6596.58-64

Object Management Group (2013). Business Process Model and Notation. *About the business process model and notation specification version 2.0.1*. https://www.omg.org/spec/BPMN/2.0.1/PDF (Accessed on 7 October 2024).

Osti, F., Ceruti, A., Liverani, A., & Caligiana, G. (2017). Semi-automatic design for disassembly strategy planning: An augmented reality approach. *Procedia Manufacturing*, *11*. https://doi.org/10.1016/j.promfg.2017.07.279

Paprocki, M. (2016). Modeling of stages of machining process planning aided by CAPP systems. *Mechanik*, *12*. https://doi.org/10.17814/mechanik.2016.12.536

Paprocki, M. (2017a). Aspekty rozwoju wyrobu i organizacji przedsiębiorstw w środowisku rozproszonym. In R. Knosala (Ed.), *Innowacje w zarządzaniu i inżynierii produkcji, Oficyna Wydawnicza Polskiego Towarzystwa Zarządzania Produkcją*, *1*, 76–87.

Paprocki, M. (2017b). Ekologiczne aspekty projektowania wyrobów. In R. Salerno-Kochan & M. Wojnarowska (Eds.), *Ecological, economic, legal and marketing aspects of products' quality, Polskie Towarzystwo Towaroznawcze*, 52–64.

Puglieri, F. N., & Ometto, A. R. (2011). Environmental and operational analysis of ecodesign methods based on QFD and FMEA. *Glocalized Solutions for Sustainability in Manufacturing – Proceedings of the 18th CIRP International Conference on Life Cycle Engineering*. https://doi.org/10.1007/978-3-642-19692-8_25

Puglieri, F. N., Ometto, A. R., Salvador, R., Barros, M. V., Piekarski, C. M., Rodrigues, I. M., & Netto, O. D. (2020). An environmental and operational analysis of quality function deployment-based methods. *Sustainability (Switzerland)*, *12*(8). https://doi.org/10.3390/SU12083486

Ragaert, K., Huysveld, S., Vyncke, G., Hubo, S., Veelaert, L., Dewulf, J., & Du Bois, E. (2020). Design from recycling: A complex mixed plastic waste case study. *Resources, Conservation and Recycling, 155*. https://doi.org/10.1016/j.resconrec.2019.104646

Realyvásquez-Vargas, A., Robles-Heredia, R., García-Alcaraz, J. L., & Díaz-Reza, J. R. (2023). Reliability tests as a strategy for the sustainability of products and production processes—A case study. *Mathematics, 11*(1). https://doi.org/10.3390/math11010208

Rosen, M. A., & Kishawy, H. A. (2012). Sustainable manufacturing and design: Concepts, practices and needs. *Sustainability, 4*(2). https://doi.org/10.3390/su4020154

Schischke, K., Hagelüken, M., & Steffenhagen, G. (2005). *An introduction to ecodesign strategies–why, what and how?* Fraunhofer IZM, Berlin, Germany.

Selvaraj, P., Radhakrishnan, P., & Adithan, M. (2009). An integrated approach to design for manufacturing and assembly based on reduction of product development time and cost. *International Journal of Advanced Manufacturing Technology, 42*(1–2). https://doi.org/10.1007/s00170-008-1580-8

Shayan, N. F., Mohabbati-Kalejahi, N., Alavi, S., & Zahed, M. A. (2022). Sustainable Development Goals (SDGs) as a framework for Corporate Social Responsibility (CSR). *Sustainability (Switzerland), 14*(3). https://doi.org/10.3390/su14031222

Shu, C., Zhou, K. Z., Xiao, Y., & Gao, S. (2016). How green management influences product innovation in China: The role of institutional benefits. *Journal of Business Ethics, 133*(3). https://doi.org/10.1007/s10551-014-2401-7

Silver, B. (2011). *BPMN Method and Style, 2nd Edition, with BPMN Implementer's Guide: A structured approach for business process modeling and implementation using BPMN 2.0.* Cody-Cassidy Press, Aptos.

Singh, S. P., Ekanem, E., Wakefield, T., & Comer, S. (2003). Emerging importance of bio-based products and bio-energy in the U.S. economy: Information dissemination and training of students. *International Food and Agribusiness Management Review, 5*(3), 14.

Spangenberg, J. H., Fuad-Luke, A., & Blincoe, K. (2010). Design for Sustainability (DfS): The interface of sustainable production and consumption. *Journal of Cleaner Production, 18*(15). https://doi.org/10.1016/j.jclepro.2010.06.002

Subramoniam, R., Huisingh, D., & Chinnam, R. B. (2010). Aftermarket remanufacturing strategic planning decision-making framework: Theory & practice. *Journal of Cleaner Production, 18*(16–17). https://doi.org/10.1016/j.jclepro.2010.07.022

Suresh, P., Ramabalan, S., & Natarajan, U. (2015). An integrated approach for the sustainable development of an automotive component using CAD/CAE, DFE and DFMA concept. *Applied Mechanics and Materials, 766–767*. https://doi.org/10.4028/www.scientific.net/amm.766-767.1009

Suresh, P., Ramabalan, S., & Natarajan, U. (2016). Integration of DFE and DFMA for the sustainable development of an automotive component. *International Journal of Sustainable Engineering, 9*(2). https://doi.org/10.1080/19397038.2015.1096313

Taylor-Smith, K. (2021). *How is 3D printing a sustainable manufacturing method?* AZO Materials.

Telenko, C., O'Rourke, J. M., Seepersad, C. C., & Webber, M. E. (2016). A compilation of design for environment guidelines. *Journal of Mechanical Design, 138*(3). https://doi.org/10.1115/1.4032095

United Nations Environment Programme, & International Resource Panel. (2018). *Redefining value: The manufacturing revolution – Remanufacturing, refurbishment, repair and direct reuse in the circular economy.*

Wang, Y., Wang, Z., Li, B., & Cheng, Y. (2023). The choice of subsidy policy for incentivizing product design for environment. *Computers and Industrial Engineering*, *175*. https://doi.org/10.1016/j.cie.2022.108883

Yadav, S., Bhudhiraja, S., & Gupta, M. (2021). Corporate social responsibility-the reflex of science and sustainability. *European Journal of Molecular & Clinical Medicine*, *7*, 6222–6233.

Yuan, Z., Sun, C., & Wang, Y. (2018). Design for manufacture and assembly-oriented parametric design of prefabricated buildings. *Automation in Construction*, *88*. https://doi.org/10.1016/j.autcon.2017.12.021

Zarębska, J. (2017). Gospodarka o obiegu zamkniętym drogą do zrównoważonego rozwoju. *Systemy Wspomagania w Inżynierii Produkcji*, *6*(7), 286–295.

Zhang, Z., Liao, H., Chang, J., & Al-barakati, A. (2019). Green-building-material supplier selection with a rough-set-enhanced quality-function deployment. *Sustainability (Switzerland)*, *11*(24). https://doi.org/10.3390/su11247153

4

THE ROLE OF LIFE CYCLE ASSESSMENT IN THE DESIGN AND DEVELOPMENT OF SUSTAINABLE PRODUCTS

Marcin Rychwalski and Tomasz Nitkiewicz

4.1 Introduction

Life cycle assessment (LCA) is a practical environmental management method that evaluates the environmental impact of a product or system throughout its entire life cycle. This method takes into account all stages in the cycle, from the sourcing of raw materials to the disposal of waste, and everything in between. The full life cycle is often referred to as a "from the cradle to the grave" cycle. LCA is a comprehensive approach that considers the potential side effects and environmental impact at every stage of the product's life (Statement from the Coalition for Higher Ambition on the EU 2030 Climate Target, 2020).[1]

In the context of the European Union's obligations to minimize the environmental impact of industry, the LCA method is a useful tool for measuring implementation. The LCA method covers the entire product life cycle, which is its key advantage. At each stage of this cycle, LCA helps examine the impact that a product has on the environment, including on humans, the natural environment, on global warming, and natural resources. The LCA method enables companies to be aware of and identify the interdependence between human activity and the natural environment. It is also an important source of information when designing, optimizing, and improving a product's features. It allows an organization to compare a product's impact on the natural environment and, consequently, to improve the condition and quality of that product (Masson-Delmotte et al., 2021).

The LCA method is not a new concept. It was first mentioned back in 1969 when Harold Smith presented his research on the production of various types of energy through selected chemical processes at the World Energy Conference. Coca-Cola was one of the first companies to show an interest in using these analyses in practice. On the other hand, the theoretical basis of LCA methodology was first

DOI: 10.4324/9781032710693-5

established in 1990 during a conference in Vermont. Today, there is considerable interest in LCA, and its results are in great demand due to the growing ecological awareness of society and the environmental regulations introduced as a result. This trend is visible almost all over the world. The LCA is now defined as a technique aimed at assessing environmental threats connected with the system, product, or operation. It identifies and quantifies materials and energy consumed as well as waste released into the environment. It also assesses the impact of these factors on the environment, covering the entire life cycle of a product as well as activities ranging from the extraction and processing of mineral raw materials, the manufacturing, distribution, use, reuse maintenance, and recycling of products, as well as final development and transportation. LCA studies the environmental impact of the product system on the ecosystem, human health, and the resources used. The LCA guidelines are regulated by the ISO 14000 series standards (Marsmann, 2000), in particular *PN-EN ISO 14040 (ISO 14040:2006(En), Environmental Management, Life Cycle Assessment, Principles and Framework,* 2006) and 14044 standards (*ISO 14044:2006(En), Environmental Management, Life Cycle Assessment, Requirements and Guidelines,* 2006).

The ISO 14040 standard sets out the LCA method for assessing the ecological aspects and potential impact of a product. It involves analyzing the inventory of ecologically significant input and output factors, assessing the potential environmental impact of input and output factors, interpreting the results of the analysis of the inventory and impact phases in relation to the objective research. The nature and potential of a product's environmental impact must be determined as objectively and repeatably as possible, which is ensured by the use of LCA procedures. This impact is the result of the interaction of many factors with complex, multi-criteria connections. It is difficult to predict the final environmental effect of the production, use, and post-use disposal of a product (Goedkoop et al., 2016).

Environmental management at the enterprise level encompasses various aspects related to the development, implementation, and achievement of environmental goals. The environmental management system covers such areas as organizational structure, responsibilities, operational practices, procedures, processes, and resources. The environmental management model consists of five basic elements: commitments and policy; planning; implementation; measurement and evaluation; and review and improvement. Product issues already occur at the stage when environmental policy is formulated. During the planning stage as well as at later stages, the identification, measurement, and assessment of environmental aspects are taken into account. The environmental aspect concerns the interactions between the activities of an organization and its products, on the one hand, and the environment on the other. This usually involves taking into account a company's impact on the environment, but the reverse impact is also possible. Any change, whether favorable or unfavorable, caused partly or entirely by the organization's activities, products, or services, is regarded as an environmental impact (Rychwalski, 2020).

4.2 Life cycle assessment rules

LCA studies can be carried out with different degrees of detail, based on the recipient's needs and the availability of input data. There are three basic variants of LCA, one of which is the conceptual variant (LCA screening), which is frequently applied within an organization. This variant is often preferred when the focus is on the speed of analysis (limited time) or a low budget. This variant takes into account estimated secondary data from existing databases or statistical studies. A sensitivity analysis is recommended so as to verify the actual impact of the obtained results on key analysis issues. The development time for this variant ranges from a few days to approximately one month.

When conducting analyses that require more detailed information, a simplified LCA is often used in decision-making processes related to product development and communication strategies. To ensure accuracy, data obtained from existing databases should be supplemented with current data from the literature and primary data from suppliers, manufacturers, or other participants in the product chain, including direct interviews or measurements. It is essential to perform a sensitivity analysis in case there is a need to correct major assumptions. The development time for this type of analysis can last from several weeks up to several months.

To ensure that a comprehensive study best reflects the object being analyzed, the detailed variant of LCA (detailed LCA) is used. This method is also employed when conducting comparative tests of products. Detailed primary data from direct measurements, analyses, and interviews are gathered and supplemented with current data from the literature to ensure statistical quality. Despite the variant of LCA applied, it is a decision-driven assessment and is heavily dependent on the competences and perhaps also intentions of the evaluator or its commissioner (Nitkiewicz & Cappelletti, 2022).

Determining the purpose of LCA is a crucial step in any analysis. The decisions made when setting the research goal are essential as they define the geographical, temporal, and spatial scope of the analysis, as well as the level of detail. This stage is of great importance as it sets the course for all other LCA activities. Identifying the stakeholders to whom the research results will be addressed is also a critical aspect that determines how the study will be carried out. This practice aligns with the amended PN-EN ISO 14040:2016 standard, which emphasizes the importance of taking into account stakeholder relationships. LCA serves as a decision support tool in the enterprise, and thus the interested parties are vital stakeholders who influence the analysis's scope, form, and level of detail. However, due to the diverse ways in which analyses are conducted as well as their purposes, comparing them with other studies can be challenging. The arrangements made at this stage are mapped in the product's system input and output inventory model (life cycle inventory [LCI]), which constitutes the next phase of the LCA. The ISO 14040 standard specifies the principles for determining the purpose of the research, which emphasizes the importance of providing a precise definition of the intended use of the analysis, the reasons for conducting the research, and the target audience for the results. Therefore, it is essential to describe the problems raised accurately and comprehensively so as to best explain the analyses performed (ISO 14044:2006(En), *Environmental Management, Life Cycle Assessment,* Requirements and Guidelines, 2006).

Determining the scope of the research requires taking into consideration various components in the study, which should be organized according to the order in which they were created. The LCI stage requires a clear description of the functions of material-energy sets of connected unit processes or the smallest parts of the product for which data is collected (product system functions). The study's quantitative effect creates a reference unit called the functional unit. The product system's comprehensive study covers all connections related to it, and not just the product itself. Therefore, the system's boundaries should indicate the interface between the product system and the environment or other product systems. The system interface areas determine which unit processes should be included in the LCA. It is recommended that the system inputs included in the analysis be elementary streams (materials or energy taken from the environment without prior processing). The criteria used to define system boundaries should be identified and justified within the scope of the definition and be consistent with the purpose of the research. Additionally, the documentation should include procedures for assigning the input or output streams of a unit process to the tested system, referred to as allocation procedures. The types and methodology of impact assessment and the interpretation method should also be included in the study's documentation. Finally, to ensure a correct analysis, the data quality requirements should be met, which means that the following analysis inputs must be carefully documented:

- the time interval,
- the geographical and technological area,
- the original, complete and representative data,
- the logicality and repeatability of the method,
- the source data and their representativeness, and
- the uncertainty of the information.

The documentation for such a study should include findings that outline the assumptions and limitations of the research. If a critical review is to be conducted, the methodology used to carry it out should also be described, along with the type and format of the report required to communicate the research results. One of the primary objectives of determining the scope of the study is to ensure that the analysis performed is detailed, comprehensive, and thorough enough to meet the stated goal. This principle applies equally to the comparative testing of product systems, where all differences between systems must be documented.

Life cycle analysis entails assigning environmental impacts to relevant inputs, thereby aiding our understanding of the environmental interventions that have taken place. This encompasses the stages of classification, characterization, and weighing. In impact analysis, the transparency of the documentation is critical, as it enables verification of the assessment's objectivity and accuracy. It is recommended that the LCA reflects the results of any sensitivity analysis conducted. The study report should cover all aspects of the analysis, including an additional report if the research was conducted for a third party. A critical aspect of life cycle analysis is verification of the data in terms of their reliability, completeness, and statistical accuracy. Various types of sensitivity analysis are acceptable as methods of data verification (Pereira et al., 2014).

Defining the functional unit is a crucial step in conducting a life cycle analysis. The functional unit can be a product, a process, or a part thereof that is analyzed. The functional unit need not be reduced to the product itself, instead, it can be a combination of various processes or elements that create a product system that is evaluated as a whole. The selection of a functional unit can be justified by conducting a sensitivity analysis in both quantitative and qualitative terms (Guince, 2002a).

4.3   LCA in sustainable product design

LCA is an essential tool for assessing the sustainability of a product. By taking into account the environmental impact of a product throughout its entire lifecycle, it is possible to create a balanced product (Trojanowski, 2017). LCA functions allow an organization to identify key impacts, assess secondary impacts, and identify areas for improvement. Thanks to these functions the method can be used to optimize environmental aspects of a product (Rychwalski, 2020).

LCA can be used both to create new products and improve existing ones. LCA is often used to optimize the environmental impacts of the design, a process known as Ecodesign. To maximize its positive impact, it is important to measure and continuously monitor it throughout the product's life cycle. To do this accurately, all stages of a product's creation must be examined, including the initial identification of a need, as well as when coming up with design ideas and when simulating various product variants, production techniques, and processes. A full and comprehensive assessment of a product's environmental impact can only be made after examining all these stages (Saouter et al., 2020).

The LCA method is an interdisciplinary approach that aims to understand the complex interactions between human activities, including industrial processes and natural systems. These interactions result in environmental impacts that are diverse and multi-dimensional, and, as a consequence, require suitable methods to ensure an accurate description and analysis. The LCA approach presents these complex levels of influence as three distinct spheres.

The first sphere, called the technosphere, deals with the modeling of technical systems such as production and transportation. Forecasts made for this sphere are characterized by negligible uncertainty as long as all measurements are verifiable and repeatable.

The second sphere, known as the ecosphere, focuses on such environmental effects as emissions. Measuring the ecosphere is complex and subject to uncertainties due to the changing nature of the environmental conditions.

The third sphere involves subjective resources and the participation of values in weighing impact categories. This sphere is known as the sphere of values, and it plays an important role when choosing the procedure and time horizon for the LCA analysis.

By means of this three-sphere approach, the LCA method provides a comprehensive understanding of the changes occurring in the environment and their impacts caused by the studied phenomena. This approach allows for the most accurate and repeatable description of any environmental changes, which is essential for making informed decisions and developing sustainable solutions (Rychwalski, 2020).

4.4   Relationships between the method and EU initiatives

The LCA method is a crucial tool for measuring the application of the European Green Deal, a package of policy initiatives that aim to transform the European Union (EU) into a modern, competitive, and ecologically sustainable economy. The ultimate goal of the Green Deal is to achieve climate neutrality by 2050.

These initiatives highlight the need for a comprehensive, cross-sectoral approach that encompasses several policy areas, including the climate, the environment, energy, transport, industry, agriculture, and sustainable finance. The package includes a variety of closely related initiatives that aim to achieve the overarching climate goal (Fetting, 2020).

The European Green Deal was launched in December 2019 by the Commission, and this was noted by the European Council at its December meeting. The Green Deal aims to promote sustainable development and combat climate change through various initiatives. These initiatives include significant changes in the functioning of businesses and the aspirations of Member States, which are of great importance for the environment. One such initiative is the Coalition for Climate Ambition 2050, which was adopted as a declaration by the European Council in 2019. It calls on its members, including businesses, investors, local and regional authorities, as well as non-governmental organizations, to commit to reducing greenhouse gas emissions by 55% or even 65% by 2030. To achieve this, a European climate review of the policies of EU Member States was also carried out, and this was included in a document presented in November 2019. The document analyzed potential policy impacts on climate neutrality goals.

There is growing interest in environmental issues, sustainable development, and the global climate crisis. This is due to increased awareness and government policies, as well as stakeholder and investor pressure. The Corporate Sustainability Reporting Directive (CSRD; Voss, 2022) came into effect in January 2023, replacing the Non-Financial Reporting Directive (NFRD Voss, 2014), which was adopted in 2014 and came into force in 2017. The new directive has expanded the scope of reporting requirements and introduced more detailed and ambitious reporting requirements for companies covered by the directive, including by expanding the group of these companies.

The CSRD applies to all companies listed on a regulated market in the EU, with the exception of micro-enterprises. It also covers all large companies that exceed two of the following three criteria according to 'Directive 2013/34/EU' (2013) on accounting: a workforce of more than 250 employees, a net turnover exceeding EUR 40 million, and a balance sheet total exceeding EUR 20 million.

Additionally, the CSRD indirectly applies to non-EU companies operating in the EU with the following parameters: a net turnover generated in the EU exceeding EUR 150 million and with a subsidiary in the EU covered by the CSRD, or a net turnover generated in the EU exceeding EUR 150 million and possessing a branch in the EU with an annual net turnover of more than EUR 40 million.

The new regulations introduced in the field of environmental reporting are designed to enhance transparency and consistency in the reporting of sustainable development by businesses. The directive aims to standardize the rules around sustainability reporting by companies in the EU, so as to ensure more accessible and transparent information for investors, consumers, employees, and other stakeholders. The CSRD applies to large public and private companies that have a significant social and environmental impact and requires them to provide minimum sustainability reporting standards, including social, environmental, and corporate governance information.

The CSRD requires companies to report their sustainability strategies, goals, activities, and results. The directive was developed following extensive public consultations, taking into account the views of various stakeholders. It is designed to be in harmony with other reporting regulations such as TCFD (Task Force on Climate-related Financial Disclosures) and SFDR (*Sustainable Finance Disclosures Regulation – European Commission,* 2019). CSRD will be implemented gradually, starting with public companies and then extended to cover other categories of enterprises. The directive emphasizes the importance of uniform sustainability reporting standards, such as Global Reporting Initiative (GRI Standards, 2021) or SASB standards (SASB® Standards, 2022). The implementation of CSRD is expected to raise public awareness, reduce business risks, and promote a long-term sustainable business strategy. It is also intended as a further step toward transforming the EU economy into a more sustainable and climate-resilient one. Companies that comply with the directive can enjoy greater stakeholder trust and be better equipped to face the challenges of sustainable development.

4.5   Environmental reporting initiatives

In the case of Poland, there are currently no specific regulations at the national level that require companies or organizations to report their carbon footprint. However, there are certain initiatives and regulations that, in some way, require reporting of greenhouse gas emissions. One such initiative is the European Union Emissions Trading System (EU ETS). Companies in sectors such as energy, industry, and aviation, which are covered by EU ETS, are required to monitor and report their greenhouse gas emissions. They must also purchase an appropriate number of emission allowances. Additionally, Poland, like other EU Member States, is required to prepare and implement the National Energy and Climate Plan, which includes targets for reducing greenhouse gas emissions. Although not directly related to reporting, achieving these goals may require monitoring emissions and taking action to reduce them.

Some business sectors may include initiatives and industry standards that encourage or require participants to report their carbon footprint. For instance, large companies operating in the food or industrial sectors may voluntarily participate in carbon footprint reporting programs.

Organizations that operate in Poland have the option of adopting international standards for environmental management and sustainability reporting. These standards include ISO 14064 standards, which apply to the management of greenhouse gas emissions, and the GRI (ISO 14064-1, 2018).

It is worth noting that there are currently no specific regulations at the national level. However, the changing global situation and growing awareness of climate change may lead to the introduction of stricter carbon footprint reporting requirements in the future.

It is important to note that the final details regarding CSDR carbon footprint reporting requirements in Poland have already been introduced and from 2024 on the following groups of companies should follow the regulations concerning non-financial reporting, including an organizational CF report (Voss, 2022).

4.6   Modern guidelines for LCA

The requirements for conducting an LCA may change over time due to changes in regulations, industry standards, and public expectations. In the field of LCA, updates and new methods are constantly being introduced in environmental databases used to assess environmental damage.

In response to the increasing interest in sustainable development, LCA can take into account a broader range of factors, including various social and economic aspects of sustainability, rather than solely focusing on the environmental impact. New requirements may encourage the inclusion of technological innovations in LCA, such as renewable technologies, recycling, or energy efficiency, which can have a significant impact on assessment results.

Some countries may introduce laws regulating carbon footprint reporting and LCA, especially in sectors with a high environmental impact, for example, the chemical industry or food production. In the electronics and automotive industries, among others, we may see an increase in demand for access to product life cycle information so that consumers can make more informed choices.

New LCA requirements may also be driven by a growing interest in sustainable development, ensuring transparency in supply chains, and corporate social responsibility. Companies are, therefore, encouraged or required to conduct more comprehensive LCAs of their products and processes (Abdi et al., 2023).

4.7   Updated versions of selected environmental databases

4.7.1   Ecoinvent database

Version 3.10 of the Ecoinvent database has been updated with new and revised data from various sectors, including agriculture, construction, chemicals, electricity, forestry and timber, fuels, metals, packaging, pulp and paper, and waste. The new version also includes updated impact assessment methods and classifications.

The agricultural update includes comprehensive data on new crop production in two major agricultural producing countries – Australia and the United States. For Australia, the update provides accurate data on crops such as barley grain, corn grain, oat grain, and wheat grain from new regions, which improves the constitution of their agricultural sector. The update for the United States includes data for corn, sweet corn, potatoes, soybeans, and various field operations from new states, thereby expanding the scope of information. Additionally, the update includes average European data for flax production and scutching. The database quality for certain agricultural products has been improved by reviewing fertilizer inputs for lentil and pea production in Canada.

The Ecoinvent database has brought significant improvements to the construction sector with the addition of fresh data and updates. New information has been included on the production of clinker and various types of cement for Tunisia, along with their corresponding market activities. The Swiss market database has been updated and outdated activities have been removed, making it compatible with current construction materials production, and now includes updated data on clinker and cement production, including Portland, CEM II/A and CEM II/B, along with their corresponding occupation markets. Extensive data sets have been introduced for the Ecuador region covering the production of building materials such as mud bricks, clay bricks, clinker, cement, concrete blocks, and clay roof tiles, along with their corresponding market activities.

The ecoinvent v3.10 update includes major improvements in the chemicals sector. The update aims to provide better representation of data regarding basic chemical precursors and their derivatives, such as short-chain alkenes (ethylene, propylene, butene, and butadiene), monocyclic aromatics (benzene, toluene, and xylenes [p-, o-, mixed]), ethylene oxide, and ethylene glycol. Industry data for European conditions was sourced from Plastics Europe.

Swiss Federal Laboratories for Materials Science and Technology (Empa) has provided three new datasets on bamboo forestry, bamboo pole production, and flattened bamboo production in Ecuador for the forestry and timber sector.

The oil and gas sector's geographical coverage has been significantly expanded. The latest version, 3.9.1, includes oil and gas production datasets from 41 different geographic areas, which represents an increase from the previous 27. This update has extended coverage to over 96% of global oil production and over 98% of natural gas production. By including data from countries such as Australia, Oman, and Turkmenistan, which are key contributors to the world's natural gas supply, especially in Asian economies, we have greatly improved sector representation.

Data related to thermal spraying techniques, such as atmospheric plasma spray, high-speed oxyfuel, and cold spray, have been added to the product and metal processing groups. Furthermore, the prices of rare earth oxides have been updated.

The pulp and paper sector has been expanded to include three new datasets that are associated with the production of beverage cartons. Additionally, the latest study conducted by the European Federation of Corrugated Board Producers (FEFCO) and Cepi Container Board (CCB) has resulted in updated datasets for corrugated board production in the European region.

There have been some updates in the waste sector focused on breaking down over 450 solid waste treatment datasets. This makes it possible to see the entire supply chain of processing activities separated into different data sets. All by-products are reported transparently, and emissions are appropriately allocated to the relevant activities that generate them. Additionally, local waste transport distances are included. Furthermore, the geographic scope of the sector has been expanded by adding data sets on solid waste treatment in other countries.

Just like with every release, we have meticulously reviewed the existing supply chains in our database to ensure that they accurately reflect the current state of global and regional product flows. This has led to updates in the connections between demand and supply activities, as well as to the introduction of new regional market data sets.

In Ecoinvent version 3, all products and basic mass exchangers are defined by at least six properties: dry mass, wet mass, water in wet mass, water content, fossil, and non-fossil carbon content. Moreover, each product in the database has a price that can be utilized for, among other purposes, economic allocation (FitzGerald, 2023).

4.7.2 Agri-footprint database

Since its release in 2014, various organizations have used the agri-footprint database for different applications, such as product LCA, hotspot analysis, carbon footprint measurements, labeling systems, and the implementation of sustainability-related software and services. Blonk has developed the Agricultural Footprint Database to provide accurate and consistent LCI data to support sustainable development in the agriculture and food sectors. It is a leading database and is used in agricultural footprint calculations.

The Agricultural Footprint Database includes data on food, feed, and agricultural intermediates. It provides transparency and enables key actors in the food system to reduce their impact on the environment.

The food industry, LCA community, scientific community, and government institutions have all generally accepted the Agricultural Footprint. Furthermore, many processes are modeled on the European Commission's Product Environmental Footprint (PEF) guidelines. The database was created on the basis of the LCA ReCiPe Midpoint H method (version 1.07, 2016), and it offers three allocation approaches: mass allocation, energy allocation, and economic allocation.

With each release, the Agricultural Footprint has improved in data quality and agricultural coverage. Currently, it contains approximately 5,000 products and processes. The basis for the Agricultural Footprint is statistics, scientific literature, other databases, and industry data combined with Blonk's expert modeling (Blonk et al., 2022).

4.7.3   Agrybalyse database

Agribalyse is a research project that has been supporting the transformation of the food industry since 2010. The program developed methodologies and reference data to assess the environmental impact of agricultural and food products, which has become necessary due to climate change and the ecological crisis.

In order to promote sustainable food practices, AGRIBALYSE® offers a dashboard of robust data on the environmental impact of agricultural and food products, along with methodologies. These are based on the LCA method, which provides indicators of the environmental impact of products throughout all stages of production, from field to fork. Various environmental issues are taken into account, such as climate, water, air, soil, and more. Depending on the context of use, additional indicators may be required to complement AGRIBALYSE® indicators, such as indicators for biodiversity, animal welfare, nutritional quality, and others.

Agribalyse data is based on research and the contributions of a network of experts, making it a reliable source of information for those interested in making informed decisions regarding the environmental impact of agricultural and food products.

AGRIBALYSE® is an open-source database that contains information on 2,500 food products and 200 agricultural raw materials consumed and produced in France, respectively. The database offers a standardized approach for analyzing the environmental impact of agricultural and food products. It also provides a platform for research aimed at improving the methodology and data, along with a network of experts helping explain and disseminate the work. Additionally, it offers documentation, training, technical days, and a dynamic of continuous improvement. The database is regularly updated and validated to ensure its quality and transparency. It is a public environmental database that publishes all indicators of the environmental impact of agricultural and food products for research and innovation purposes when designing food products. It is also an excellent source of information for consumers (Auberger et al., 2022).

4.7.4   Carbon footprint dedicated datasets

Since the LCA approach is used to help develop much simpler CF methodology, its LCI databases might be used for these calculations. Nevertheless, there is a growing number of dedicated CF datasets that usually comprise conversion factors from a range of industry wide activities, including energy use, water consumption, waste disposal, recycling, transport activities, as well as materials or market products. The list of dedicated CF datasets includes, but is not limited to, global sets as CCaLC, DEFRA, EXIOBASE, GREET, and the IPCC Emissions Factor Database, as well as localized sets like U.S. EPA Supply Chain Emission Factors for U.S. Industries and Commodities, AusLCI, and 3EID (Sotos, 2015). It seems that CF datasets have an advantage over regular LCI databases since access to them is usually free and are updated more often (Nitkiewicz et al., 2023). On the other hand, CF datasets are often limited to specific flows and substances and in order to cover complex product assessments the user very often needs different sources. This strategy could be a bit risky, due to the different standards and procedures involved, as well as the data background used for developing datasets.

Databases are useful in determining substance quantities and processes. However, when assessing environmental damage, appropriate methods are necessary, and these are divided into geographical areas.

4.8   Methods in product balancing – carbon footprint

The carbon footprint is an essential part of the CSRD directive and EPD environmental reporting. It acts as an environmental indicator and helps assess the organization, its products, and processes by means of LCAs. The principles determining a carbon footprint may vary based on the context and standards used. Generally, however, the process comprises some key principles and steps. The first step involves establishing the scope of the analysis, which could be a product, process, service, or organization. It is important to set the system boundaries, which means deciding what will be included in the analysis (e.g., emissions from the production, transport, use, and disposal of the product) (European Investment Bank, 2023). After that, data on greenhouse gas emissions connected with the selected scope of analysis needs to be collected. This data can be obtained from various sources such as emissions databases, company reports, product LCAs, and field studies. Data for different greenhouse gases (e.g., CO_2, methane, nitrous oxide) are converted into CO_2 equivalents so as to consider their different heat absorption capacities (emission factors or Global Warming Potential [GWP] factors).

Once all the data has been collected, the emissions are converted into an equivalent amount of CO_2, and a life cycle analysis is conducted. The results of the analysis are then verified, and a report is prepared that includes information on the calculated carbon footprint, the methodology used to conduct the analysis, and possible actions recommended to reduce it.

It is essential that the scope of the analysis is defined, and all significant greenhouse gas emissions associated with the facility under examination are taken into account. The carbon footprinting process is a critical step in reducing the environmental impact of products and processes (Vallero, 2019).

4.9   Methods for assessing environmental damage and the different categories of these methods

LCA is a useful tool that connects a product system with its various environmental impacts. It involves the use of impact assessment methods that help determine the environmental pressures exerted by a product system. These methods are diverse and updated regularly to address various environmental concerns. In this chapter, we will discuss some of these methods, which are characterized by their relevance to contemporary environmental problems (Goedkoop et al., 2016).

There are currently six categories of methods available in LCA software. The European category covers comprehensive life cycle impact assessment (LCIA) methods that primarily focus on Europe and are useful when conducting LCA studies in Europe. The global category includes comprehensive LCIA methods that have a global reach and are ideal for studies involving a global value chain. The North American category includes methods developed for the North American region. The Single-Issue category includes methods that focus on one single indicator or environmental impact in a single area, except for those focusing on water. The Water Footprint category includes methods for assessing only water-related impacts. The Superseded category includes methods that are archived and no longer promoted. We strongly discourage users from selecting these options. However, they are still stored and distributed to current users who can apply them for their own needs.

4.9.1   The Intergovernmental Panel on Climate Change global warming potential

The Intergovernmental Panel on Climate Change (IPCC) has developed an Environmental Damage Assessment Method called GWP. It was first introduced within the framework of the EPD 20182 method used by the IPCC to evaluate the impact of various greenhouse gases on global warming. The GWP method measures the greenhouse potential of the gas in comparison to carbon dioxide (CO_2) over a 100-year period. GWP 100a 2021 is an improved and updated version of the previous method that takes into account the latest scientific research and data on greenhouse gas emissions and their impact on climate change (Masson-Delmotte et al., 2021).

The GWP method 100a is widely applied in carbon footprint assessments of products, processes, services, and organizations. When an organization reports its greenhouse gas emissions, GWP 100a values are used to convert the emissions of different gases into an equivalent amount of CO_2. This makes it easier to compare their impact on global warming (Iordan et al., 2016).

GWP values are updated regularly because scientific research on the impact of greenhouse gases on the climate is constantly evolving. Therefore, the GWP 100a values used in the IPCC method have been modified compared to its EPD 2018 predecessor, and they may be further updated in the future to better reflect the latest data and scientific developments in the field of climate change.

4.9.2   CML-IA method

In 2001, a team of researchers led by the Center for Environmental Sciences at Leiden University proposed a set of categories and methods for analyzing impacts during the evaluation stage. They came up with an impact assessment method called CML-IA, which follows the middle approach. This methodology offers normalization, but neither weighing nor addition. It comes in two versions: one with ten "mandatory" impact categories and an extended version with "all impact categories", which includes additional impacts for different time frames.

The CML Guide, compiled by Guinee (2002b), provides a comprehensive list of impact assessment categories that are divided into three groups, namely mandatory impact categories, additional impact categories, and other impact categories. If multiple methods are available for mandatory impact categories, the baseline indicator is chosen based on the principle of the best available practice.

4.9.2.1   Characteristics of the impact categories

Abiotic resource depletion is a category of environmental impact that focuses on protecting human health, well-being, and the ecosystem. This impact category concerns the extraction of minerals and fossil fuels and is measured using an abiotic depletion factor (ADF) for each type of extraction. The ADF is determined according to the concentration reserves and degree of deaccumulation and is expressed as kg antimony equivalent/kg extraction. The geographical scope of this indicator is global. Climate change is another factor that affects ecosystem health, human health, and materials. Climate change is caused by the emission of greenhouse gases into the air and can have adverse effects on the environment and human health (Huijbregts et al., 2003).

4.10   Conclusions

As time passes, the methods for assessing the impact of human activities on the environment need to be updated and adapted to changing needs and regulations. New approaches and interpretations are being introduced to keep up with the evolving understanding of environmental issues. Modern LCA methods are designed to take into account dynamic factors such as climate change, resource depletion, and technological progress, which had not been suitably accounted for in previous static approaches. In recent years, there has been a growing interest in integrating LCA methods with the circular economy model, which helps us better understand and assess the impact of economic processes on the environment and resource management. The use of advanced IT tools, including simulation and modeling software, makes it possible to provide a more accurate analysis and forecast the impact of human activities on the environment. With the availability and quality of data from different regions of the world, it is now possible to adopt a more global approach to assessing the life cycle impact of products and processes. In addition, modern LCA methods are increasingly taking into account the social and economic aspects of this impact, allowing for a more comprehensive assessment of sustainable development. All these changes reflect the need for more precise and comprehensive tools to assess the impact of human activities on the environment, and to consider a wide range of factors in decision-making. The LCA method has various practical applications when determining the environmental footprint of a product, in particular its carbon footprint. These tasks have been incorporated in particular into the provisions of the CSRD directive and have also appeared previously in the concept of Sustainable Development.

Acknowledgement

The publication presents the result of the Project no 060/ZJE/2023/POT financed from the subsidy granted to The Krakow University of Economics.

Note

1 About 484 European companies have committed to or have already set science-based targets (SBTs) and are already raising EU ambitions regarding the climate. One hundred and fifty nine leading European companies have committed to aligning their businesses with a 1.5°C future through the 'Business Ambition for 1.5°C' initiative launched 'SBTs' campaign.

References

Abdi, M., Rohani, A., Soheilifard, F., & Taki, M. (2023). Energy optimization and its effects on the environmental repercussions of honey production. *Environmental and Sustainability Indicators, 17.* https://doi.org/10.1016/j.indic.2023.100230

Auberger, J., Ayari, N., Ceccaldi, M., Cornelus, M., & Geneste, C. (2022). *Agribalyse Version 3.1 Change Report.* www.ademe.fr/mediatheque

Blonk, H., Tyszler, M., van Paassen, M., Braconi, N., van Rijn, J., & Draijer, N. (2022). *Agri-Footprint 6 Methodology Report.* https://blonksustainability.nl/tools/agri-footprint (accessed 26.10.2022)

Directive 2013/34/EU. (2013). https://eur-lex.europa.eu/eli/dir/2013/34/oj.

European Investment Bank. (2023). *EIB Project Carbon Footprint Methodologies Methodologies for the Assessment of Project Greenhouse Gas Emissions and Emission Variations,* Luxembourg. https://www.eib.org/en/publications/20220215-eib-project-carbon-footprint-methodologies#:~:text=The%20EIB%20Project%20Carbon%20Footprint%20Methodologies%20contain%20the.

Fetting, C. (2020). "The European Green Deal", Europeans Sustainable Development Network ESDN Report, December 2020, ESDN Office, Vienna. Goedkoop, M., Oele, M., Leijting, J., Ponsioen, T., & Meijer, E. (2016). *Introduction to LCA with SimaPro Title: Introduction to LCA with SimaPro.* www.pre-sustainability.com

GRI Standards. (2021). https://www.globalreporting.org/how-to-use-the-gri-standards/gri-standards-english-language/

Guinee, J. B. (2002a). Handbook on life cycle assessment operational guide to the ISO standards. *The International Journal of Life Cycle Assessment, 7*(5), 311–313. https://doi.org/10.1007/BF02978897

Guinee, J. B. (2002b). Handbook on life cycle assessment operational guide to the ISO standards. *The International Journal of Life Cycle Assessment, 7*(5). https://doi.org/10.1007/BF02978897

Huijbregts, M. A. J., Breedveld, L., Huppes, G., De Koning, A., Van Oers, L., & Suh, S. (2003). Normalisation figures for environmental life-cycle assessment: The Netherlands (1997/1998), Western Europe (1995) and the world (1990 and 1995). *Journal of Cleaner Production, 11*(7), 737–748. https://doi.org/10.1016/S0959-6526(02)00132-4

Iordan, C., Lausselet, C., & Cherubini, F. (2016). Life-cycle assessment of a biogas power plant with application of different climate metrics and inclusion of near-term climate forcers. *Journal of Environmental Management, 184,* 517–527. https://doi.org/10.1016/j.jenvman.2016.10.030

ISO 14040:2006(en), Environmental Management, Life Cycle Assessment, Principles and Framework (2006). https://www.iso.org/obp/ui/en/#iso:std:iso:14040:ed-2:v1:en

ISO 14044:2006(en), Environmental Management, Life Cycle Assessment, Requirements and Guidelines (2006). https://www.iso.org/obp/ui/en/#iso:std:38498:en

ISO 14064-1. (2018). *ISO 14064-1:2018(en), Greenhouse Gases — Part 1: Specification with Guidance at the Organization Level for Quantification and Reporting of Greenhouse Gas Emissions and Removals.* https://www.iso.org/obp/ui/en/#iso:std:iso:14064:-1:ed-2:v1:en

Marsmann, M. (2000). The ISO 14040 family. *The International Journal of Life Cycle Assessment, 5*(6), 317–318. https://doi.org/10.1007/bf02978664

Masson-Delmotte, A., Zhai, V., P., Pirani, A., Connors, S. L., Péan, C., Berger, S., Caud, N., Chen, Y., Goldfarb, L., Gomis, M. I., Huang, M., Leitzell, K., Lonnoy, E., Matthews, J. B. R., Maycock, T. K., Waterfield, T., Yelekçi, O., Yu, R., & Zhou, B. (2021). *The Physical Science Basis Climate Change 2021 Working Group I Contribution to the Sixth Assessment Report of the Intergovernmental Panel on Climate Change.* Cambridge University Press. https://doi.org/10.1017/9781009157896

Nitkiewicz, T., & Cappelletti, G. M. (2022). Verification of circular economy solutions and sustainability of products with life cycle assessment. *Sustainable Products in the Circular Economy: Impact on Business and Society,* 36–52. https://doi.org/10.4324/9781003179788-3

Nitkiewicz, T., Wiszumirska, K., & Rychwalski, M. (2023). *Circular Solutions for Food Packaging. Innovative Coated Paper Packaging and Its Carbon Footprint.* https://doi.org/10.29119/1641-3466.2023.186.36

Pereira, L. G., Dias, M. O. S., Junqueira, T. L., Pavanello, L. G., Chagas, M. F., Cavalett, O., Maciel Filho, R., & Bonomi, A. (2014). Butanol production in a sugarcane biorefinery using ethanol as feedstock. Part II: Integration to a second generation sugarcane distillery. *Chemical Engineering Research and Design*, *92*(8), 1452–1462. https://doi.org/10.1016/j.cherd.2014.04.032

FitzGerald, D., Bourgault, G., Vadenbo, C., Sonderegger, T., Symeonidis, A., Fazio, S., Mutel, C., Müller, J., Dellenbach, D., Stoikou, N., Baumann, D., Clementi, M., Ioannou, I., Cirone, F., Superti, V., Beckert, P., Treichel, A., Kaarlela, O., Kunde, S., Valsasina, L., Moreno Ruiz, E. (2023). *Documentation of Changes Implemented in the Ecoinvent Database v3.10 (2023.12.14),* Zürich https://19913970.fs1.hubspotusercontent-na1.net/hubfs/19913970/Knowledge%20Base/Database/Releases/Change%20Report%20v3.10%20-%2020231214.pdf#:~:text=This%20report%20covers%20the%20changes%20to%20the%20ecoinvent%20database

Rychwalski, M. (2020). *Analiza cyklu życia w zarządzaniu produktem.* https://libra.ibuk.pl/reader/analiza-cyklu-zycia-w-zarzadzaniu-produktem-marcin-rychwalski-248786

Saouter, E., Biganzoli, F., Ceriani, L., Versteeg, D., Crenna, E., Zampori, L., Sala, S., & Pant, R. (2020). Environmental footprint: Update of life cycle impact assessment methods – Ecotoxicity freshwater, human toxicity cancer, and non-cancer. In *JRC Technical Reports* (Issue December). http://doi.org//10.2760/300987

SASB® Standards (2022). https://sasb.ifrs.org/standards/download/#company-search-form

Sotos, M. (2015). *An amendment to the GHG Protocol Corporate Standard GHG Protocol Scope 2 Guidance.* World Resources Institute, USA. https://ghgprotocol.org/sites/default/files/2023-03/Scope%202%20Guidance.pdf

Statement from the Coalition for Higher Ambition on the EU 2030 Climate Target (2020). https://caneurope.org/content/uploads/2020/10/Final-version-_-October-HAC-letter-on-EU-2030-Climate-Target.pdf

Sustainable Finance Disclosures Regulation – European Commission (2019). https://finance.ec.europa.eu/regulation-and-supervision/financial-services-legislation/implementing-and-delegated-acts/sustainable-finance-disclosures-regulation_en

Trojanowski, T. (2017). Projektowanie zrównoważonych produktów. *Zeszyty Naukowe Politechniki Śląskiej Organizacja i Zarządzanie*, *100*(1972), 1–10. www.unesco.pl

Vallero, D. A. (2019). Air pollution calculations: Quantifying pollutant formation, transport, transformation, fate and risks. *Air Pollution Calculations: Quantifying Pollutant Formation, Transport, Transformation, Fate and Risks*, 1–556. https://doi.org/10.1016/C2017-0-02742-8

Voss, W. G. (2014). *The European Union'S 2014 Non-Financial Reporting Directive: Mandatory Ex Post Disclosure-But Does it Need Improvement?* https://hal.science/hal-02562673

Voss, W. G. (2022). *Directive -2022/2464- EN – CSRD Directive - EUR-Lex.* https://eur-lex.europa.eu/legal-content/EN/TXT/?uri=CELEX:32022L2464

5

SUSTAINABLE PRODUCTION

Alina Matuszak-Flejszman

5.1   Introduction

Managing operations in an environmentally and socially responsible manner – "sustainable manufacturing" – is no longer excellent to have but a business imperative. Companies worldwide face increased materials, energy, and compliance costs coupled with higher expectations of customers, investors, and local communities (OECD, 2011). Sustainable production is part of Sustainable Development Goal 12, entitled Sustainable Consumption and Production, which promotes the efficient use of energy and other resources, sustainable infrastructure, access to essential services, decent jobs, including in the environmental sector, and a better quality of life. In line with this goal, actions should be taken to produce and consume goods to ensure the sustainable management of resources and their efficient use.

The European Union is carrying out particular activities in this area, promoting the implementation of a circular economy. This concept is based on keeping raw materials in circulation for as long as possible, extending the life cycle of products and reducing waste. This will improve the environmental efficiency of the production process and products throughout their life cycle and increase the demand for environmentally friendly and high-quality products and production technologies (Muradin, 2023).

It is indisputable that while industrial growth is pivotal for the progress and development of any country, it is also recognized as a significant source of pollution and resource depletion, leading to environmental degradation (Herva et al., 2011). Therefore, for industries, the challenge is to balance economic progress and societal advancement with environmental conservation. While sectors have begun to grasp the importance of sustainable development, they may need help understanding how to operationalize this concept. Industries must incorporate environmental

DOI: 10.4324/9781032710693-6

aspects into the production process, product design, and value chain management to prevent unsustainable resource utilization and adverse environmental impacts. Initiatives such as industrial ecology, cleaner production, and design for the environment are a result of such realization (Herva et al., 2011).

The leadership of many organizations has already started taking essential steps towards green growth – ensuring their economic and environmental development. Their pioneering experiences demonstrate that improving the environment goes hand in hand with achieving profits and increased competitiveness.

However, in addition to environmental activities, the organization should implement appropriate social activities and activities related to ensuring proper corporate governance to raise the awareness and involvement of all employees and people working under the organization's supervision.

Many other tools can be used in organizations, not only in the field of environmental management but also in the field of social responsibility and risk management. These tools may be based on international or national standards, programmes or guidelines that can be used in whole or in part to ensure adequate supervision of sustainable production. Another essential element of supervision over sustainable production activities is the assessment of the effects. For this purpose, many tools and guidelines related to key performance indicator (KPI) analysis can be used. However, non-financial reporting guidelines for business information on sustainability can also be used. Entrepreneurs are increasingly interested in developing and publishing the non-financial effects of their activities in the form of ESG (Environment, Social responsibility, corporate Governance) indicators.

ESG refers to how corporations and investors integrate environmental, social, and governance issues into their activities. ESG explicitly covers matters relating to organizational governance (Gilan et al., 2021), including sustainable production management. To identify business risks and increase investor and consumer confidence, non-financial disclosure is critical to managing the shift towards a sustainable global economy by combining social justice and environmental protection. In this context, disclosing non-financial information helps measure, monitor, and manage business performance and, therefore, sustainability accounting (Vukić et al., 2018).

Original equipment manufacturing firms must maintain sustainable production and adapt their managerial responses to changing environments to sustain their competitive edge. To provide such a response, a firm must identify the sustainable production indicators (SPIs) for overhauling the production process to achieve the firm's goal of waste elimination and reduce the impact on the environment. The SPIs enable the firm's continuous improvement in its environmental impact with great emphasis on green product development in a competitive and sustainable market. The primary cause of the global environment's continued deterioration is the unsustainable consumption and production pattern, especially in industrialized nations such as Taiwan (Tseng, 2013).

The challenges entrepreneurs face in sustainably implementing production processes will translate into many benefits resulting from activities carried out sustainably and supervised, considering good management practices while meeting legal requirements and assuring stakeholders that the company's activities align with the Sustainable Development Goal 12. Sustainable production should not pose a risk to future generations and should not be carried out at the expense of future generations.

5.2   The essence and concept of sustainable production

There is no clear definition of the term "sustainable production." "Sustainable manufacturing" is the formal name for an exciting new way of doing business and creating value. It is behind many eco-friendly products and processes that are in demand and celebrated worldwide today. Businesses already engage in initiatives and innovations that help create a healthier environment, increase their competitive advantage, reduce risk, build trust, stimulate investment, attract customers, and generate profits (OECD et al., 2011). US Department of Commerce's Sustainable Manufacturing Initiative sums it up as: "The creation of manufactured products that use processes that minimize negative environmental impacts, conserve energy and natural resources, are safe for employees, communities, and consumers and are economically sound." The Lowell Center for Sustainable Production defined sustainable production as "the creation of goods and services using processes and systems that do not pollute the environment." This definition also added that sustainable production should take into account "energy conservation and natural resources." Moreover, sustainable production should be "economically viable, safe and healthy for workers, communities and consumers, and socially and creatively fulfilling for all working people".

Sustainable production is a term defining a specific model of conducting business in the manufacturing industry, one of the effects of which is minimizing the impact on the environment. However, minimizing the negative environmental impact is not the only effect. It should be emphasized that sustainable production plays an increasingly important role in shaping the future of manufacturing enterprises because it not only helps protect the environment but also improves the company's reputation and brings numerous financial benefits. Thus, the term sustainable production refers to several areas of production:

- aimed at producing products and services at an appropriate level that meet the needs and expectations of customers,
- supervised in such a way as to minimize the negative impact on the environment and prevent pollution,
- taking into account the needs and expectations of employees and other people working under the organization's supervision,

- supervised using quality, environmental and safety management programmes and systems,
- conducted effectively and efficiently.

Sustainable production minimizes the diverse business risks inherent in any manufacturing operation while maximizing the new opportunities that arise from improving your processes and products.

A company's reputation is related to improving its social impact and ensuring appropriate corporate governance. Sustainable production should involve changing the way of production. Production efficiency should be increased using fewer raw materials, thus resulting in lower costs and limited environmental impact. Particular changes in the field of sustainable production are needed in the food, construction, and transport sectors because it is in these industries that the most significant impact of production on the environment is recorded (Baraniewicz-Kotasińska, 2022).

As shown, there needs to be a clear definition of sustainable production. However, it is a production that considers environmental, economic, and social aspects when implementing processes and achieving specific goals. Below are three-dimensional aspects of sustainable production (Table 5.1).

Considering these three aspects in sustainable production makes achieving specific goals for sustainable development possible. These goals can be implemented

TABLE 5.1 Three-dimensional aspects of sustainable production

Economy	i.) Society	ii.) Environment
Contributing to local economy	Complying with the law	Efficiently using of energy
Investing in infrastructures	Good working conditions	Efficiently using of resources
Combating bribery and corruption	Ensuring products safety	Using environmentally sound materials
Creating jobs	Treating suppliers fairly	Protecting biodiversity
Driving innovations	Respecting human rights	Minimising waste
Paying tax responsibly	Good community relations	Minimising energy consumption
Generating sales and profits	Employees' right to information and consultation	Using renewable energy sources
Increasing energy and material efficiency in a sustainable way	Occupational health and safety	Reducing greenhouse gas emissions
Lower costs through environmental solutions	Universal ecological education	Minimising use of hazardous substances
Industry 4.0 & 5.0	Improving the quality of life	Minimising emissions

through tasks that can be implemented to many extents and in other ways depending on the management's commitment and a sustainable approach. Sustainable production will therefore depend on sustainable management of the organization, which can be achieved through:

- setting goals for minimizing negative impact on the environment, preventing pollution, and positively influencing the improvement of quality of life;
- introducing the principles of sustainable business management through effective use of natural resources;
- reducing losses in the production and distribution process;
- ensuring the sustainable and environmental management of chemicals and all waste throughout their entire life cycle;
- reducing the level of waste generation through prevention, reduction, recycling and reuse;
- implementing sustainable development practices and including information on this subject in its periodic reports;
- promoting sustainable procurement practices;
- providing access to relevant information and raising stakeholders' awareness regarding sustainable development and a lifestyle in harmony with nature.

Sustainable manufacturing practices have been defined primarily from an environmental perspective, aiming to minimize the environmental impact of manufacturing activities while optimizing the organization's production efficiency (Nordin et al., 2014). Accordingly, sustainable manufacturing practices are understood as activities, initiatives, and techniques that positively impact a company's environmental, social, or economic performance by helping to control or mitigate the impact of an organization's activities across the triple bottom line.

5.3 Principles of sustainable production

The LCSP has developed principles that reflect the main aspects of sustainable production. These include (Alayon et al., 2017):

1 Products and packaging are designed to be safe and ecologically sound throughout their life cycles, and services are also designed to be safe and environmentally sound.
2 Wastes and ecologically incompatible by-products are reduced, eliminated, or recycled.
3 Energy and materials are conserved, and the energy and materials used are most appropriate for the desired ends.
4 Chemical substances, physical agents, and conditions that present hazards to human health or the environment are eliminated.

5 Workplaces and technologies are designed to minimize or eliminate chemical, ergonomic, and physical hazards.
6 Management is committed to an open, participatory process of continuous evaluation and improvement focused on the firm's long-term economic performance.
7 Work is organized to conserve and enhance the efficiency and creativity of employees.
8 The security and well-being of all employees and the continuous development of their talents and capacities are priorities.
9 The communities around workplaces are respected and enhanced economically, socially, culturally, and physically; equity and fairness are promoted.

These principles reflect the aspects of sustainable production, from designing products sustainably to planning the production process, production, distribution, and disposal of these products. They also consider issues related to the sustainable packaging and distribution of these products. It should be emphasized that sustainable production includes activities aimed at reducing energy consumption, reducing the consumption of non-renewable resources, water and material consumption, and ensuring the use of materials in the production process that are safe for the environment, employees, customers, and other stakeholders. These principles also include eliminating chemicals or agents threatening human health aimed at employees and customers. These principles highlight the social factor in ensuring appropriate working conditions, improving the participatory management style, promoting stakeholder involvement in decision-making, and increasing customer satisfaction.

In line with these aspects, management should engage in an open, participatory process of continuous evaluation of processes, products, and employees, focusing on the organization's long-term economic performance. Sustainable production means that management maximizes profits, minimizes the negative environmental impact, and actively engages in social initiatives. In the context of manufacturing companies, production processes are optimized in terms of economic, environmental, and social efficiency aimed at achieving profits and minimizing the negative impact of the organization's activities on the environment and society.

They are implementing sustainable production principles in as many enterprises as possible belonging to the European manufacturing industry, which brings numerous social and environmental benefits. It improves their functioning and thus increases their economic efficiency.

5.4   Management of sustainable production

An organization wishing to manage itself sustainably can use many management and process improvement tools and rely on strategies, concepts, programmes, and management principles that may consider issues related to sustainable development,

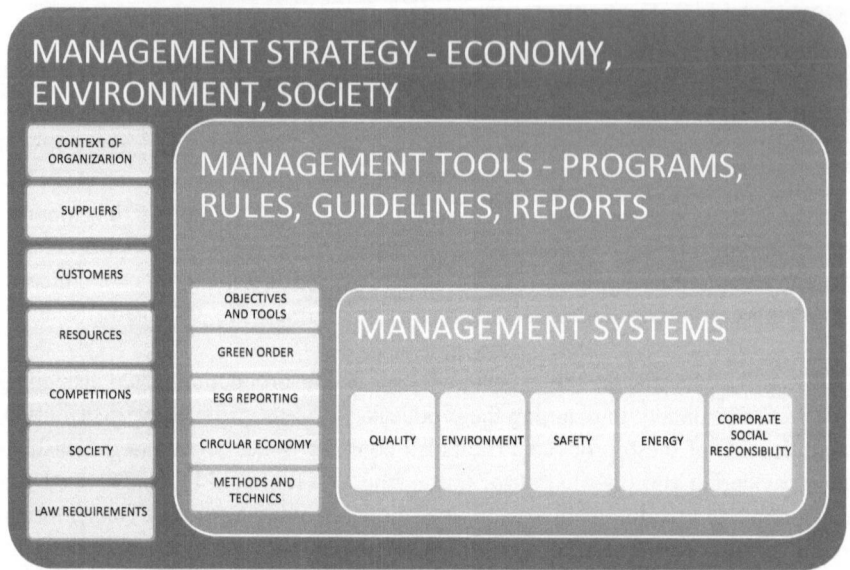

FIGURE 5.1 Management strategy for sustainable organization management.

focused on the economy, environment, and society. In Figure 5.1, possible actions of top management in sustainable organization management are presented.

Both concepts, strategies, and management programmes are comprehensive plans and initiatives that organizations implement to manage their activities sustainably and effectively. However, sustainable management tools help the organization's management effectively manage quality, safety, environmental, and social aspects (Matuszak-Flejszman, 2023). To implement activities resulting from the sustainable management strategy, the organization's management can use many standardized or otherwise formalized tools, which are very helpful in implementing sustainable production activities. Sustainable management tools may be quality management systems (e.g., ISO 9001, ISO 22000, ISO 13485, IATF 16949, ISO/TS 22163-IRIS, AS/EN 9100), environmental management systems (ISO 14001, Eco-Management and Audit Scheme [EMAS], Cleaner Production), human resources management (HRMS, ISO/TR 30406), safety management systems (SMS, ISO 27001, ISO 45001), corporate social responsibility (e.g., SA 8000, ISO 26000, BSCI, ISO 20400), or sustainable development of society (ISO 37101).

An essential element is the implementation of sustainable production based on internal factors shaping the sustainable management of the organization:

- leadership and commitment of top management,
- customer orientation,
- risk-based approach,
- evidence-based decision-making,

- employee engagement,
- process approach,
- improvement,
- managing relationships with stakeholders.

5.4.1 The leadership and commitment of top management

The leadership and commitment of top management, including strategic and operational planning, is one of the most critical aspects of sustainable production. Sustainable production management should be a priority for the organization's top management and be cascaded to subsequent levels in the organizational structure. The organization's management is responsible for strategic planning in the aspect of sustainable development and creating a sustainable management policy that considers not only aspects related to ensuring an appropriate level of quality but also environmental policy and policy focused on corporate social responsibility. Therefore, Lean Green can be introduced in the organization, which is an operating strategy that uses selected Lean tools and techniques to eliminate eco-waste, such as unnecessary consumption of energy, water, raw materials, and materials, while also reducing emissions of pollutants and the amount of waste generated (Bachorz, 2023). However, we must remember the third pillar – the impact on society. Figure 5.2 shows the direction of the organization's strategy aimed at sustainable production.

To this end, the organization's management should:

- include in its strategy issues related to minimizing the negative impact on the environment resulting from the organization's activities and social responsibility,

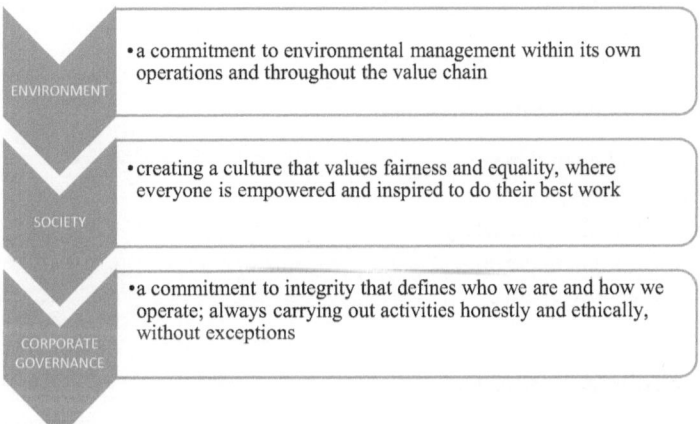

FIGURE 5.2 Directions of the organization's strategy aimed at sustainable production.

taking into account good practices of corporate governance and cooperation with stakeholders;
- communicate its mission, vision, strategy, policies, and processes throughout the organization and create and maintain shared values of sustainability, as well as integrity and ethical behaviour at all levels of the organization;
- ensure an appropriate culture of trust and integrity and universal support for commitment to quality, environmental protection, and safety;
- provide employees and people working under the organization's supervision with the required resources, training and the opportunity to obtain additional qualifications, and inspire and encourage them to act sustainably and responsibly;
- appreciate the contribution of employees and people working under the organization's supervision;
- ethically conduct business and ensure corporate culture (anti-bribery and corruption, animal welfare).

As a result of implementing sustainable production activities, the organization's management can set the following goals:

- in the environment area:
  - sustainable customer development,
  - sustainable resource management,
  - energy and emissions management,
  - responsible supply chain;

- in the area of society:
  - talent and culture,
  - diversity, equality, and inclusion,
  - employees of tomorrow,
  - occupational health and safety;

- in the area of corporate governance:
  - ethics and compliance,
  - cyber security,
  - product quality and safety,
  - corporate risk management,
  - corporate governance.

5.4.2   Customer orientation

Customer orientation is another important aspect of sustainable management that translates into sustainable production. Sustainable production should consider the needs and expectations of the customer and other interested parties, not only to ensure the appropriate quality of products but also to address issues regarding the impact of production and its effects on the environment and society. For this purpose, it is necessary to identify both direct and indirect customers and determine their current and future needs related to product quality, the impact of these products on the environment and social safety, and sustainable development. As part of sustainable production, all customer needs and expectations should be included in the organization's goals, communicated to all employees and implemented through tasks and activities at individual workstations. Customer relationships should be active.

5.4.3   The risk-based approach

The risk-based approach is a control approach that considers risks and opportunities. It significantly impacts the production process's planning, implementation, supervision, monitoring, and improvement. A risk-based approach to sustainable production enables the organization's management to identify factors that may cause deviations from planned results regarding processes, quality requirements, and environmental, social, and legal aspects and introduce preventive supervision measures to minimize adverse effects and maximize emerging opportunities. The organization's management should have internal control and risk management systems implemented. Risk management includes, among others, defining the context to identify, analyse, assess, and reduce risks about the organization's activities, processes, functions or products, risk monitoring, reporting, and recording results. All risks related to quality assurance, environmental impact, employee safety, and social impact should be considered in this area. Risk management is more effective when it is carried out by a team in which individual members have various qualifications and experience in this field. It is also worth seeking the opinion of stakeholders.

5.4.4   Evidence-based decision-making

Evidence-based decision-making is another important factor in sustainable production. Monitoring and measuring processes and other activities in terms of quality, environment, and safety, as well as analysing the results and assessing them against specific requirements, should constitute the basis for decision-making by the organization's management. Monitoring and measurement data and its analysis and evaluation should be shared with the right people and should be accurate, reliable, and secure. Based on this information, management should make decisions and actions balanced with experience and intuition.

Appropriate human resources management to ensure employee involvement in sustainable production is critical. To this end, management should communicate to employees how significant their contribution to sustainable production is, including the quality of processes or products implemented and the impact on the environment, safety, and society. Therefore, management should assign them appropriate roles, powers, and responsibilities in this area and appropriately motivate them to carry out tasks according to the established assumptions, taking into account the needs and expectations of stakeholders. Each employee should be treated equally, and as part of sustainable management, management should guarantee equal opportunities for everyone (equal pay, development opportunities, integration of excluded people, prevention of violence). Each employee and person working under the supervision of the organization should be provided with appropriate working and employment conditions (working time, freedom of association, employees' right to information and consultation, occupational health, and safety). Sustainable production must also ensure respect for fundamental human rights.

5.4.5   Process approach

The process approach involves controlling interrelated or interacting processes and activities to achieve intended results. It enables understanding and consistently fulfilling requirements, considering processes in terms of added value, achieving the effectiveness of implemented processes, and improving them based on data and information.

One condition for sustainable production should be a shift from long-term benefits and their consequences to short-term profits. Therefore, the appropriate design, planning, implementation, and optimization of sustainable processes, considering the organization's environmental policy and methods of introducing employees to appropriate practices, plays an important role.

The production process aims to produce at the lowest possible costs while maintaining the highest dimensional accuracy. At the same time, the process should be ecological, clean, and sustainable. Sustainable production is an activity that does not threaten future generations and is not carried out at the expense of future generations. Clean production cannot mean increased financial investments (Krolczyk et al., 2019).

As part of sustainable management, sustainable production should be implemented using the Deming cycle – Plan-Do-Check-Act (PDCA), considering the stages of planning activities, implementing activities, assessing activities, and working towards continuous improvement.

Management should define the processes necessary to achieve the established goals and establish authority, responsibility, and accountability for managing these processes. As part of the process approach, it is essential to determine the interdependence between these processes and how introducing improvements in one process may affect other processes by identifying environmental aspects, quality, and safety threats and risks. Processes should be adequately supervised to ensure the appropriate quality of products, employee safety, and minimization of the negative impact on the environment. To ensure adequate supervision of operational activities, it is necessary to plan and supervise the processes needed to meet the requirements for the delivery of products and services. Therefore, product quality, safety, environmental, and social impact criteria should be defined. Next, appropriate, sustainable design and development of products and services must be ensured. When designing processes, one should consider the possibility of using various innovations that play an essential role in improving quality, minimizing the negative impact of the company's activities on the environment, enhancing safety, and improving the use of products.

Process supervision should also include appropriate sourcing of raw materials, materials, semi-finished products and services, and ensuring supervision of outside processes, products, and services.

In sustainable production, an essential aspect is the implementation of activities into processes. As part of these activities, you can include Industry 4.0 or 5.0 solutions, that is, introduce intelligent machines into production. The introduction of appropriate sustainable, human- and environmental-friendly technologies into processes promotes the development of entrepreneurship. Such technologies protect the safety of users and the environment, generate less pollution, use smaller amounts of resources more rationally, ensure the reuse of created products and waste, and ensure the neutralization of generated waste in a more rational way than the technologies to which they are alternatives. Moreover, introducing intelligent machines into production significantly impacts the profitability of operations, the environment, and the safety of employees and company infrastructure. Organizational management is increasingly aware that when developing its business, it must invest in sustainable management, taking into account automation and intelligent solutions.

The production and service delivery process should be supervised sustainably. Management should also ensure appropriate risk management (quality, environmental, social, and safety) to achieve appropriate effects from these processes' activities. Identifying, understanding, and sustainably managing interconnected processes increases the effectiveness and efficiency of the organization's management activities.

5.4.6   Improvement

Improvement in sustainable production should occur by promoting the establishment of goals for ensuring the appropriate quality of products and services, minimizing the negative impact on the environment, and ensuring proper working conditions at all levels of the organization. Employees should have knowledge, be competent and participate in many projects aimed at improving activities in the field of sustainable production. For this purpose, you can use various tools to solve problems, improve activities, and introduce green lean management to reduce the negative environmental impact. This approach is often called Eco Lean Management. It is a holistic approach combining classic lean management goals with environmental optimization. Its goal is continuously improving economic and environmental performance through small steps (Kaizen) – low-cost organizational improvements (Schutzbach et al., 2022). In addition to the seven types of waste, these activities can include waste management as an unavoidable side effect of the production process and environmental waste, which is considered any unnecessary use of resources or substances released into the air, water or soil that may harm human health or the environment of the production process. As a result of these activities, production will become even more sustainable. The results of improvement projects should be assessed in terms of sustainability and, after meeting specific criteria, implemented in production processes.

5.4.7 Relationships with stakeholders

Another critical factor influencing sustainable production is stakeholder relationship management. It is essential because it affects the organization's relationships with suppliers and other interested parties such as partners, customers, investors, employees, and society. Therefore, relationships with these stakeholders need to be identified and prioritized. Important stakeholders influence an organization's performance. It is essential to establish relationships between them that balance short-term benefits with long-term considerations. Therefore, the impact of these stakeholders on the organization's performance must be optimized by quickly responding to the opportunities and constraints associated with each stakeholder.

Moreover, stakeholders should know the goals resulting from sustainable production and jointly adapt to them by implementing cooperation criteria. By sharing information, resources, and competencies, added value is created for all interested parties. Measuring performance and providing feedback to interested parties is good practice for good stakeholder relations and contributes to improving sustainable production. Moreover, in managing relationships, it is essential to properly manage risks related to quality, environmental impact, and safety. A well-managed supply chain will ensure a stable flow of products and services.

The above factors are sets of subsequent components (constituting elements) that, to varying degrees, affect sustainable production, its effectiveness and, as a result, the success of sustainable management in the organization.

5.5 Assessment of the effects of sustainable production

Implementing sustainable production activities should be assessed in terms of meeting specific requirements set by top management and resulting from the organization's sustainable operating strategy. This is achieved through numerous indicators and measures that can be used to assess the effectiveness of individual processes affecting sustainable production. Thanks to their analysis and assessment, the organization's management can obtain measurable information regarding the results of its sustainable activities and the effectiveness of actions taken concerning quality, safety, environmental aspects, and stakeholder requirements. Indicators of the effects of sustainable operations can constitute KPIs for the organization for processes related to operational activities within sustainable production. Therefore, they should meet all KPI requirements. A KPI indicates results that an organization considers significant and draws attention to some aspect of operations, management, conditions, or influences. Using these indicators, the organization's management can assess the organization's effectiveness in sustainable management. It is a process that facilitates management decisions regarding performance in terms of quality, safety, and the organization's environmental and social impact by selecting indicators, collecting and analysing data, assessing information against sustainable performance criteria, reporting and communicating, and periodically reviewing and improving this process.

Unfortunately, there is no common framework for assessing sustainable production at the industrial level. Dahl (2012) observes that the steep increase in food and energy prices points to the vulnerability of global sustainability, and Bebbington et al. (2007) conclude that there is an inherent need for models, metrics, and tools to convey the extent to which anthropogenic activities are unsustainable. As a result, the ongoing debate on sustainability and the development of tools to measure its progress have taken a new urgency.

Various measures or indicators, such as those based on the Global Reporting Initiative (GRI) guidelines, the EMAS regulation requirements, and the Corporate Sustainability Reporting Directive (CSRD), can be used for this purpose.

The first step in selecting indicators is to familiarize yourself with the list of fundamental indicators that management in each organization should consider, regardless of the sector. They refer to the areas listed by the NFRD (Directive 2014/95/EU): the environment, social, labour issues, human rights, and anti-corruption. The organization's management can and should make this data public. Reporting on sustainable management allows you not only to share your achievements with your stakeholders but also to determine the organization's most significant impacts on the environment, society, and economy, as well as to organize internal processes, revise set goals, and analyse the results achieved.

An organization's most frequently chosen reporting method worldwide is a separate sustainability report. It allows you to collect all information on environmental, social, and corporate governance issues in one document. The integrated report combines elements of the financial report and sustainable development issues in one document, showing how the organization's strategy and value-building model influence the ESG activities undertaken and their results (Matuszak-Flejszman et al., 2023). For this reason, this format is more often chosen by the management of more advanced organizations with a well-developed system for managing issues related to the economy, society, and the environment. Integrated reporting is promoted by the International Integrated Reporting Council (IIRC). To facilitate comparison of reports published by organizations, ESG disclosures are encouraged by generally accepted standards and reporting frameworks, particularly GRI, IIRC, and SASB. According to the GRI guidelines, efficiency indicators are grouped into six categories. The six categories cover environment (EN1–EN30), economics (EC1–EC9), society (SO1–SO8), human rights (HR1–HR9), labour practices and decent work (LA1–LA14), and product responsibility (PR1–PR9). Directive (EU) 2022/2464 of the European Parliament and of the Council of 14 December 2022 on corporate reporting on sustainable development defines three groups: environment (E1–E5), social responsibility (S1–S4), and corporate governance (G1). In July 2023, the first set of European Sustainability Reporting Standards (ESRS) was published, including a set of mandatory indicators: sector-independent, sector-specific, and entity-specific, with 84 quantitative and qualitative disclosure requirements.

Those responsible for ESG reporting of organizations rely on various methods, approaches, and tools still being developed in national and international institutions, which influence statutory requirements and the content of voluntary reports (Kocmanova et al., 2012). Although the term ESG was introduced in 2004 by the United Nations, there still needs to be more consistency in terms of features, attributes, and standards defining individual ESG components (Billio et al., 2021). Thus, reports published by various organizations are often criticized because they must fully illustrate how financial and non-financial elements are managed to create enterprise value (Hoang, 2018). For example, Ellili (2020), Sharma et al. (2020), and Suttipun (2021) examined the scope of information reporting on environmental, social, and governance data and confirmed that, although still at a low level, the scope of information has increased in the following years. Additionally, corporate governance information constitutes the most significant part of ESG reporting, followed by social and environmental information. Moreover, several recent studies (Manita et al., 2018; Arayssi et al., 2020; Shakil, 2021; De Masi et al., 2021) have verified the impact of various corporate governance mechanisms on ESG reporting.

The topic of ESG is also reflected in the corporation's strategies, which proves the high priority given to this issue. Still, despite the high involvement of both corporations in ESG activities, a comparison of the presented indicators is not possible, even though both companies operate in the same industry. This example confirms the need to create a uniform reporting scheme for ESG indicators undertaken by the European Parliament and the Council of the European Union (Kamińska-Witkowska & Matuszak-Flejszman, 2023). However, such information would be necessary from the point of view of assessing the effectiveness of activities resulting from sustainable production.

5.6 Conclusions

Both large manufacturing enterprises, as well as small- and medium-sized enterprises, benefit from the implementation of sustainable production, including the adoption of new, sustainable operating procedures that take into account issues related to meeting the needs and expectations of customers, related to ensuring an appropriate level of quality, safety, impact on the environment, and society. It should be emphasized that as part of sustainable production, the company benefits from proper corporate governance and minimizing the adverse effects on the environment and society. When implementing sustainable production processes, company employees "want to feel good in what they do every day." They should be aware that their work contributes to sustainable development. In a sustainably managed organization, employees know and implement its goals and are aware that they are part of not only booked profits, financial growth, and revenue streams but, above all, activities aimed at preventing pollution and taking care of the well-being of stakeholders and our planet. The company's corporate social responsibility programmes, sustainable production practices, and shared quality and environmental impact projects help increase employee engagement.

The benefits for the company resulting from the use of sustainable production include:

- achieving financial benefits resulting from:
  - process optimization,
  - more efficient use of raw materials and energy,
  - increasing sales as a result of better matching the offer to the environmental requirements of an aware customer,
  - increasing efficiency by limiting the consumption of resources and the amount of waste generated, minimizing environmental fees,
  - eliminating possible problems related to the use of hazardous substances;
- introducing changes in the way of operation based on good practices in advance of legal regulations in the field of environment and safety, which avoids the need to react quickly when they enter into force,
- planning, design, development, sustainable production, delivery, and support of products and services to meet customer needs and expectations in a sustainable way;
- universal support throughout the organization of commitment to quality, environmental protection, safety, and social impact,
- improving the image and increasing the attractiveness of the organization among customers,
- better understanding by people of the goals regarding quality, minimizing the adverse effects on the environment and society of the organization, and increasing motivation to achieve them,
- providing people with the required resources, training and empowerment to act responsibly and sustainably,
- increased care for shared values and culture throughout the organization,
- increasing people's involvement in activities to improve sustainable production,
- identification of critical interested parties (such as suppliers, partners, customers, investors, employees, and society as a whole), their needs and expectations

in the areas of quality, environment, safety, and economy and their relationship with the organization,
- achieving positive relations with society near the production plant,
- access to new markets and new opportunities related to meeting specific sustainable production standards,
- measuring performance and providing stakeholders feedback on sustainable production performance to enhance improvement initiatives.

The implementation of solutions based on intelligent machines also brings particular and immediate benefits, which include:

- savings resulting from the planned consumption of raw materials, energy, and other resources, which translates into minimization of losses and better control over expenses,
- more excellent safety resulting from continuous monitoring of the production line condition by intelligent machines, which allows for quick response to possible irregularities and minimization of the risk of accidents and costly breakdowns,
- employment optimization as a result of the use of sustainable production, which does not mean the elimination of jobs, but their transformation; modern operating technologies allow employees to perform tasks in safer and more comfortable conditions, which increases productivity and also translates into employment stability,
- increased efficiency because intelligent machines operate more efficiently and precisely than traditional devices, while process automation and the ability to monitor and optimize machine operation allow for a significant increase in production efficiency,
- higher product quality because intelligent machines can control production processes in real-time, which translates into excellent product quality, while production parameters are monitored and intelligent machines can effectively respond to possible deviations, ensuring excellent results,
- environmental protection, thanks to optimized production processes and waste reduction, is possible to reduce the negative impact of industry on the environment, and modern technologies allow for more effective management of energy consumption and other media, which translates into lower emissions of pollutants.

Unfortunately, implementing sustainable production practices can also pose barriers. It may require investment in new technologies, staff training, and infrastructure changes. These upfront costs can be a problem for some businesses.

Another barrier may be insufficient employee awareness. To ensure the effectiveness of the sustainable production process, employees must be aware of their activities' impact on quality, environment, and safety and committed to implementing these activities. Here, training and education are essential to successfully implementing sustainable practices.

In many companies, the traditional mentality focused on short-term profits must be changed to a long-term perspective that considers the company's impact on the reality around us. This is why management's commitment and leadership role are essential to sustainable development.

In business management, a barrier to implementing sustainable production may be the risk of weakening reputation. An organization's top management may declare sustainable practices but cannot maintain them, which can damage the company's reputation. Therefore, top management's declarations in quality, environment, safety, and sustainability policies must be consistent with actual activities.

References

Alayon, C., Safsten, K. & Johansson, G. (2017). Conceptual sustainable production principles in practice: Do they reflect what companies do?, *Journal of Cleaner Production*, 141, 693–701, https://doi.org/10.1016/j.jclepro.2016.09.079

Arayssi, M., Jizi, M. & Tabaja, H. (2020). The impact of board composition on the level of *ESG* disclosures in GCC countries, *Sustainability Accounting, Management and Policy Journal*, 11(1), 137–161, https://doi.org/10.1108/SAMPJ-05-2018-0136

Bachorz, M. (2023). *Lean Green – zredukuj koszty odpadów, energii i wody. (Lean Green – Reduce the Costs of Waste, Energy and Water)*, https://leancenter.pl/bazawiedzy/lean-green (access: 27.03.2024).

Baraniewicz-Kotasińska, S. (2022). Zrównoważona produkcja i technologia (Sustainable production and technology), in: (ed.) Drosik, A., Heidrich, D. & Ratajczak, M., *Wprowadzenie do zrównoważonego rozwoju (Introduction to sustainable development)*, Wydawnictwo Naukowe Scholar, Warszawa, 233–242.

Bebbington, J., & Frame, B. (2007). Accounting technologies and sustainability assessment models, *Ecological Economics*, 61(2–3), 224–236. https://doi.org/10.1016/j.ecolecon.2006.10.021

Billio, M., Costola, M., Hristova, I., Latino, C. & Pelizzon, L. (2021). Inside the ESG ratings: (Dis)agreement and performance, *Corporate Social Responsibility and Environmental Management*, 28, 1426–1445, https://doi.org/10.1002/csr.2177

De Masi, S., Słomka-Gołębiowska, A., Becagli, C. & Paci, A. (2021). Toward sustainable corporate behaviour: The effect of the critical mass of female directors on environmental, social, and governance disclosure, *Business Strategy and the Environment*, 30, 1865–1878, https://doi.org/10.1002/bse.2721

Dahl, A. L. (2012). Achievements and gaps in indicators for sustainability. *Ecological Indicators*, 17, 14–19. https://doi.org/10.1016/j.ecolind.2011.04.032

Ellili, N.O.D. (2020). Environmental, social, and governance disclosure, ownership structure and cost of capital: Evidence from the UAE, *Sustainability*, 12(18), https://doi.org/10.3390/su12187706

Gilan, S.L., Koch, A. & Starks, L.T. (2021). Firms and social responsibility: A review of ESG and CSR research in corporate finance, *Journal of Corporate Finance*, 66, 101889, https://doi.org/10.1016/j.jcorpfin.2021.101889

Herva, M., Franco, A., Carrasco, E.F. & Roca, E. (2011). Review of corporate environmental indicators, *Journal of Cleaner Production*, 19(15), 1687–1699.

Hoang, T. (2018). *The Role of Integrated Reporting in Raising Environmental, Social, and Corporate Governance (ESG) Performance Awareness. Awareness of Environmental, Social and Corporate Governance (ESG) Performance*, Stakeholders, Governance and Responsibility, 31.08.2018.

Kamińska-Witkowska, A. & Matuszak-Flejszman, A. (2023). Possibility of using EMAS environmental reporting requirements for ESG reporting in selected automotive corporations, *Economics and Environment*, 2(85), 347–368, https://doi.org/10.34659/eis.2023.85.2.588

Kocmanova, A., Nemecek, P. & Docekalova, M. (2012). *Environmental, Social and Governance (ESG) Key Performance Indicators for Sustainable Reporting*, 7th International Scientific Conference "Business and Management 2012", May 10–11, Vilnius, Lithuania.

Krolczyk, G.M., Maruda, R.W., Krolczyk, J.B., Wojciechowski, S. Mia, M., Nieslony, P. & Budzik, G. (2019). Ecological trends in machining as a key factor in sustainable production – A review, *Journal of Cleaner Production*, 218, 601–615, https://doi.org/10.1016/j.jclepro.2019.02.017

Manita, R., Bruna, M.G., Dang, R. & Houanti, L.H. (2018). Board gender diversity and ESG disclosure: Evidence from the USA, *Journal of Applied Accounting Research*, 19(2), 206–224, https://doi.org/10.1108/JAAR-01-2017-0024

Matuszak-Flejszman, A. (2023). Strategie, programy i narzędzia zarządzania środowiskowego (Environmental management strategies, programs and tools), in: (ed.) Matuszak-Flejszman, A., *Zarządzanie środowiskowe (Environmental Management)*, Wydawnictwo UEP, Poznań, 45–52.

Matuszak-Flejszman, A., Łukaszewski, S. & Budna, K. (2023). Reporting sustainable development in polish commercial banks, *Engineering Management in Production and Services*, 15(3), 42–52, https://doi.org/10.2478/emj-2023-0019

Muradin, M. (2023). Zrównoważone zarządzanie środowiskiem (Sustainable environmental management), in: (ed.) Matuszak-Flejszman, A., *Zarządzanie środowiskowe (Environmental Management)*, Wydawnictwo UEP, Poznań, 13–34.

Nordin, N., Ashari, H. & Rajemi, M.F. (2014). A case study of sustainable manufacturing practices, *Journal of Advanced Management Science*, 2(1), 12–16, https://doi.org/10.12727/joams.2.1.12-16

OECD Sustainable Manufacturing Toolkit (2011). https://www.oecd.org/innovation/green/toolkit/48661768.pdf (access: 29.03.2024).

Schutzbach, M., Kiemel, S. & Miehe, R. (2022). Eco lean management – Recent progress, experiences and perspectives, *Procedia CIRP*, 107, 350–356, https://doi.org/10.1016/j.procir.2022.04.057

Shakil, M.S. (2021). Environmental, social and governance performance and financial risk: Moderating role of ESG controversies and board gender diversity, *Resources Policy*, 72, https://doi.org/10.1016/j.resourpol.2021.102144

Sharma, P., Panday, P. & Dangwal, R.C. (2020). Determinants of environmental, social and corporate governance (ESG) disclosure: A study of Indian companies, *International Journal of Disclosure and Governance*, 17(4), 208–217, https://doi.org/10.1057/s41310-020-00085-y

Suttipun, M. (2021). The influence of board composition on environmental, social and governance (ESG) disclosure of Thai listed companies, *International Journal of Disclosure and Governance*, 18(4), 1–12, https://doi.org/10.1057/s41310-021-00120-6

Tseng, M.L. (2013). Modelling sustainable production indicators with linguistic preferences, *Journal of Cleaner Production*, 40, 46–56, https://doi.org/10.1016/j.jclepro.2010.11.019

Vukić, N.M., Vuković, R. & Calace, D. (2018). Non-financial reporting as a new trend in sustainability accounting, *Journal of Accounting and Management*, 7(2), 13–26.

6

A SUSTAINABLE SUPPLY CHAIN

Artur Jachimowski and Martin Straka

6.1   Introduction

In today's globalized world, the concept of a sustainable supply chain (SSC) is of paramount importance, particularly when it concerns the production and distribution of sustainable products. A SSC takes into account and integrates various environmental, social, and economic concerns to ensure that resources are utilized efficiently while minimizing negative impacts on the environment and society. This approach not only addresses immediate environmental concerns but also fosters long-term resilience and profitability for businesses. One of the key principles of an SSC is environmental stewardship. This involves reducing carbon emissions, minimizing waste generation, and conserving natural resources throughout the entire supply chain (SC) process, from sourcing raw materials to manufacturing, distribution, and end-of-life disposal. Adopting eco-friendly practices, such as the use of renewable energy sources, implementing green transportation methods, and employing sustainable packaging solutions, can significantly reduce the carbon footprint of operations. Moreover, social responsibility is another critical aspect of an SSC. Companies are increasingly expected to uphold ethical labour practices, promote diversity and inclusion, and ensure the fair treatment of workers across their SC. This entails engaging with suppliers that adhere to labour standards, provide safe working conditions, and support community development initiatives in areas where operations are conducted. Furthermore, economic viability is fundamental to the success of an SSC. While initial investments may be required to implement sustainable practices, the long-term benefits, including savings in costs, an enhanced brand reputation, and access to new markets, far outweigh the upfront expenses. By optimizing processes, reducing waste, and fostering innovation, businesses can create value while simultaneously advancing sustainability goals.

DOI: 10.4324/9781032710693-7

Collaboration and transparency are key to building an SSC for sustainable products. Close cooperation among stakeholders, including suppliers, manufacturers, distributors, and customers, facilitates an exchange of ideas, best practices, and resources, and in this way drives continuous improvement. Transparency in operations, SC traceability, and the disclosure of sustainability performance metrics enhance accountability and trust among all parties involved.

This chapter covers various topics related to SSCs. It starts with a discussion of the concept, structure, and fundamental principles of SSCs. This chapter also explores the concepts of sustainable logistics and reverse logistics within the context of the SC. The priorities of SSC management (SSCM) are also presented, along with a discussion of sustainable procurement (SP) and purchasing. There is a particular focus on collaboration with suppliers when designing sustainable products. Finally, examples of activities and benefits resulting from the implementation of an SSC are presented. The aim of this chapter is to describe in detail an SSC and its components. In addition, this chapter discusses issues related to the development of an SSC and identify those factors that exert direct pressure on the SC.

6.2 Concept, structure, and fundamental principles of sustainable supply chains

SCs are systems of organizations, people, activities, information, and resources that operate as a network of entities engaged in the production and distribution of goods (Azzi et al., 2019; Flores-Sigüenza et al., 2021). Managing these chains poses a major challenge for many companies, especially those operating globally and facing strong competition. Supply chain management (SCM) is a conscious effort by firms to run chains in the most efficient way possible, ensuring coordination without negatively impacting product quality and customer satisfaction. The long-term success of SCs requires taking into account not only economic factors but also environmental and social issues, which necessitate adopting a sustainable approach. This involves integrating environmental, social, and economic aspects into business operations, contributing to the development of an industrial system conducive to recovery (Silva & Figueiredo, 2020), and the emergence of the sustainable supply chain (SCC) area (Barbosa-Póvoa et al., 2018). Embracing sustainable development in SCs requires making responsible decisions and building a capacity for long-term value for stakeholders, while being aware of pressures from politics, law, consumers, and competition (Apeji & Sunmola, 2022).

Bearing in mind the above, an SSC can be described as a set of complex networked systems comprising various entities managing products, from suppliers to customers and associated returns, taking into account various social, environmental, and economic impacts (Barbosa-Póvoa, 2014). Meanwhile, another researcher defines an SSC as *a system of interconnected business activities covering the entire product lifecycle, enabling the creation of value for all stakeholders, while also ensuring commercial success that contributes to the growth of social prosperity and improvement of the environment* (Krzysztofek, 2014).

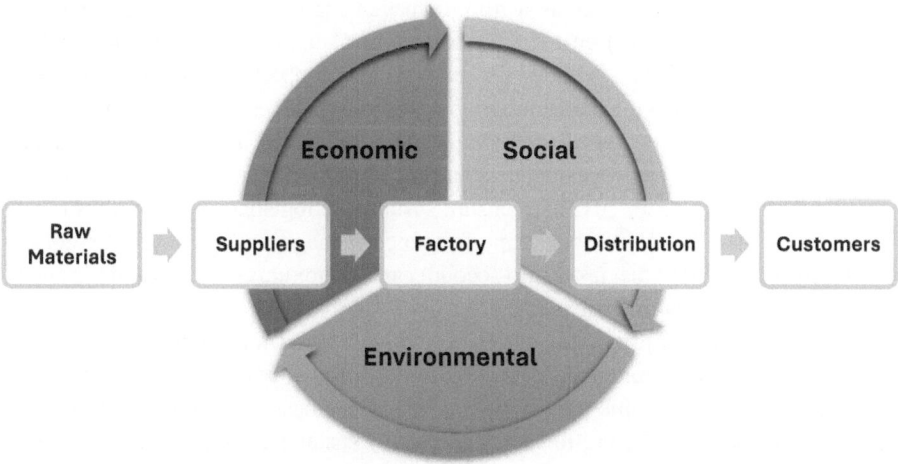

FIGURE 6.1   Sustainable supply chain diagram.

Source: Own work.

The structure of an SSC is characterized by a holistic approach that considers environmental, social, and economic factors at every stage of the SC process (de Almeida Santos et al., 2020). This stands in contrast to traditional SCs (Gao et al., 2020), which primarily focus on cost efficiency and profit maximization. In light of the above observations, an SSC development framework that takes into account the flow of resources has been proposed (Figure 6.1). It is also worth noting that at its core, the structure of an SSC comprises interconnected entities, including suppliers, manufacturers, distributors, retailers, and customers. These entities collaborate (Chen et al., 2017; Mehdikhani & Valmohammadi, 2019) to ensure that products are sourced, produced, transported, and disposed of in a manner that minimizes negative environmental and social impacts while maximizing economic benefits. This often requires collaboration and cooperation between stakeholders to implement sustainable practices such as responsible sourcing, energy efficiency, waste reduction, and fair labour practices.

One of the key elements in an SSC is reverse logistics, which involves efficiently managing the backward flow of raw materials, products, and information from the point of consumption back to the point of origin, aiming to recover value or ensure proper disposal (Ganesh Kumar & Ashlin Nimo, 2020; Mahadevan, 2019). It is often perceived as a cost-generating area, in that it seeks to reduce financial losses and minimize environmental impact (Shekarian et al., 2022). However, due to the growing importance of sustainable development, reverse logistics is increasingly becoming a way to achieve environmental goals while generating financial profits. The integration of sustainable concepts in reverse logistics affects the economic, social, and environmental aspects of businesses (Letunovska et al., 2023).

On the other hand, logistics as a flow management system (material, financial, informational, and other) ensures the balancing of market needs (Straka, 2019) and sustainable logistics includes the analysis and support of sustainable orders, transportation, packaging, distribution, and reverse logistics, as well as the design and control of actions related to SSCs (Wang et al., 2018). Its aim is to persuade decision-makers to approve logistics initiatives which respect the environment (Björklund & Forslund, 2019). To ensure sustainable logistics, organizations must implement a sustainable logistics performance management process (Persdotter Isaksson et al., 2019) that takes into account carbon dioxide emissions and the need to create SSCs (Kaur & Singh, 2019).

The fundamental principles of SSCs revolve around three main pillars: economic viability, environmental stewardship, and social responsibility (Luthra & Mangla, 2018). Economic viability ensures that businesses remain profitable and competitive while adhering to the principles of sustainability (Esmaeilian et al., 2020; Sajjad et al., 2020). Environmental stewardship focuses on minimizing the environmental footprint of operations by reducing resource consumption, emissions, and waste generation (Krishnan et al., 2020; Zhang et al., 2023). Social responsibility entails promoting ethical labour practices, ensuring worker safety and well-being, and contributing positively to local communities (Fernando et al., 2022; Mughal et al., 2023). These pillars provide the foundation for building SCs in accordance with the principles of sustainable development.

6.3   Priorities of sustainable supply chain management

One important means of promoting sustainable development within organizations (Sánchez-Flores et al., 2020), and at the same time also an area of intense research, is SSCM. It has been a topic of interest in many recent publications (Abualigah et al., 2023; Das & Hassan, 2022; Faramarzi-Oghani et al., 2023; Heidary Dahooie et al., 2021; Islam et al., 2020; Kshetri, 2021; Leal Filho et al., 2023; Lis et al., 2020; Liu et al., 2023; Mageto, 2021; Men et al., 2023; Paul et al., 2021; Sembiring et al., 2020; Tsai et al., 2021). The idea of SSCM first emerged in 2000. However, it was only after 2004 that researchers began to show a strong interest in this area (Khan et al., 2021). It is important to point out that sustainable development is a key topic in the field of SCM, and organizations are increasingly incorporating it into their long- and short-term decision-making strategies (Esfahbodi et al., 2017; Haessler, 2020). Authors such as Carter and Rogers (2008), Seuring and Müller (2008), and Ahi and Searcy ((2013) have offered their own definitions of SSCM and underscored the importance of integrating sustainable development initiatives with SCM for the current and future development of organizations. Moreover, rapid changes in customer preferences, increased competition, and pressure from governments and other stakeholder groups have led most companies to adopt sustainable practices in their SCs (Gopal & Thakkar, 2016). Bearing in mind the above considerations, SSCM can be defined as "the management of material and information flows, as well as enterprise interaction, along the SC while taking into account all important components of sustainable development" (Seuring et al., 2008).

In view of the above, the priorities of SSCM encompass three main areas: social responsibility, environmental protection, and economic efficiency. They come down to ensuring fair working conditions, minimizing negative environmental impact, and achieving profitability and operational competitiveness. These priori ties (Table 6.1) constitute a crucial foundation for effective SCM in the spirit of sustainable development.

TABLE 6.1 The priorities of SSCM

Environmental dimension of SSCM	Social dimension of SSCM
• Using eco-friendly production methods • Implementing waste reduction measures • Ensuring pollution-free production processes • Incorporating renewable energy sources • Emphasizing material reuse • Managing defective and expired products • Choosing supply chain partners with ecological considerations in mind • Getting employees involved in environmental protection initiatives	• Implementing ethical practices for employees and contractors • Prioritizing fair employment practices within the local community • Providing equipment for hygiene and job safety • Investing in infrastructure development • Ensuring timely and lawful tax payments • Maintaining income transparency for tax purposes • Adhering to ethical business practices • Supporting poverty reduction programs • Engaging in local charity initiatives • Participating in regional development projects

Business dimension of SSCM

• Collaborating in the field of inventory and logistics management
• Utilizing information technologies to improve communication efficiency
• Establishing long-term relationships based on defined guidelines
• Sharing a clear vision of supply chain management
• Implementing the "Just in Time" concept for enhanced competitiveness
• Regularly exchanging production information through meetings such as those concerning sales and operations planning
• Introducing benchmarking and performance metrics collectively
• Standardizing quality policies for products and processes based on established guidelines
• Aligning product strategies, supply, and distribution within the overall supply chain strategy
• Sharing information about customer requirements and design plans
• Incorporating the supply chain concept into product, process, and packaging design
• Implementing procedures with the aim of gathering customer feedback during product development

Source: Author's own table based on Kot (2018).

6.4 Sustainable procurement and purchasing in the context of designing sustainable products

SP and purchasing practices have become increasingly important in today's business landscape. With growing awareness of environmental and social issues, organizations are recognizing the significance of integrating sustainability into their SCs. This entails not only taking into account the environmental impact of products but also ensuring ethical sourcing practices and supporting suppliers committed to sustainability. Collaboration with suppliers is crucial, especially in the context of designing sustainable products, as it involves working together to incorporate eco-friendly materials, reduce waste, and enhance product longevity. By prioritizing SP and fostering partnerships with suppliers, businesses can drive positive environmental and social change while also meeting consumer demand for ethically produced goods.

To encourage organizations to achieve sustainable development goals, one option involves actively enhancing their capabilities in the area of procurement, particularly through the implementation of sustainable public procurement practices. Among various SSC practices, SP motivates not only the central firm but also its supply chain entities to strive towards a common goal, which is the roadmap of sustainable development. This approach embraces the idea expressed by the United Nations (2015) that SP practices have significant potential in achieving Sustainable Development Goal 12, known as Sustainable Consumption and Production, and having a positive impact on other sustainable development goals, as all 17 sustainable development goals are interconnected (Kannan, 2021).

When the concept of sustainable development, known as the triple bottom line (TBL) approach, is incorporated into the procurement process, it is known as sustainable public procurement (Kannan, 2021; Prier et al., 2016). According to Walker and Brammer, SP *is consistent with the principles of sustainable development, such as ensuring a strong, healthy and just society, living within environmental limits, and promoting good governance* (Walker & Brammer, 2009). Among the several definitions of SP that exist (Muthugala & Nayagam, 2012; Preuss, 2009; Thomson & Jackson, 2007), the one (Defra, 2006) proposed by the UK Sustainable Procurement Task Force in the "Procuring the Future" report defines it as a process in which organizations meet their needs for goods, services, works, and utilities in a way that achieves value for money on a whole-life basis by generating benefits not only for the organization but also for society and the economy, while at the same time minimizing harm to the environment (Islam et al., 2017; Kannan, 2021; Y. Messah et al., 2023; Y. A. Messah et al., 2023). It is essential for firms to invest in sustainable practices, focusing their suppliers on the task of meeting stakeholder requirements regarding compliance, social commitments, and satisfying customer demands (Govindan et al., 2020, 2021; Niu & Mu, 2020). The decision of an organization to implement SP is largely based on stakeholder requirements/pressures and greatly depends on the sustainable development performance of their suppliers and subcontractors. Considerations regarding sustainable development issues also have significant potential to impact a firm's relationships with its higher and lower-tier supply chain members (Ageron et al., 2012). Without engaging supply chain members in the TBL concept, implementing SP practices will be challenging for the central enterprise. Therefore, a central firm that wants to implement SP practices must develop its capabilities and be aware of the preferences/engagement of its customers, suppliers, and subcontractors when it comes to implementing SP (Ghadimi et al., 2016).

Another key element in the SSC that is playing an increasingly important role in business and society is sustainable purchasing, which is widely discussed in the literature on the subject (Arora et al., 2020; Houé & Duchamp, 2021; Schulze & Bals, 2020; Teixeira et al., 2018). Introducing sustainable purchasing practices involves making decisions based on social, environmental, and economic criteria. It also requires considering the product life cycle as a whole, including production, distribution, usage, and all the way through to recycling and disposal. In this way, companies can strive to minimize the negative impact of their operations on the environment and local communities, while promoting innovation, ethical business practices, and social responsibility (Miemczyk et al., 2012). Therefore, sustainable purchasing in supply chains represents a significant step towards more sustainable and responsible business operations.

In response to the growing pressure to develop sustainable products, companies have begun to implement practices such as product life cycle optimization or environmentally friendly design principles to address sustainability issues during the process of new product development (NPD; Sroufe et al., 2000). In addition to internal actions, researchers and practitioners indicate that the supply base can also serve as an important source of innovation, which should be utilized by companies in their efforts to develop novel products and services (Genç & Di Benedetto, 2015; Mackelprang et al., 2018). Therefore, researchers examining SCM have focused on demonstrating how supplier engagement in the new product development process, both in terms of timing (Petersen et al., 2005) and scope (Schoenherr & Wagner, 2016), positively impacts outcomes in this area (Wang et al., 2021). The beneficial effects of supplier engagement in the NPD process have also been noted in the context of creating sustainable new products. As a result, supplier engagement may enhance the manufacturer's ability to introduce sustainable innovations (Adomako, 2020; Melander, 2018; Oliveira et al., 2018; Yang & Wang, 2020).

Given the increasing importance of creating environmentally friendly products, as well as the efforts of company managements to utilize both internal sustainable design practices and external suppliers in sustainable innovation initiatives, it is essential to understand how these two factors collectively impact a company's efforts to successfully develop new environmentally friendly products. Any evaluation of the success of sustainable design practices must take into account both their environmental and economic effectiveness, with the implication suggested by Wang et al. (2021) that the relationship between design practices and performance should be moderated by supplier involvement. Moreover, other research confirms that collaborating with suppliers in product development, often indicated by the level of supplier involvement in the NPD process, allows a company to leverage the experience and knowledge of suppliers, which can lead to better outcomes in the product development process (Barrane et al., 2021; DeCampos et al., 2022; Murali et al., 2023; Slot et al., 2020). This is possible because supplier involvement during the early stages of product development ensures better access to ideas, smoother communication, and a faster problem-solving process, which in turn can increase productivity and speed (Parker et al., 2008). In this context, supplier involvement has been observed both in terms of time and the scope (Wynstra et al., 2012). Therefore, supplier involvement is defined as the degree to which activities are conducted in collaboration with the supplier across all five stages (from idea generation to prototype development) of an NPD project aimed at creating an environmentally friendly product (Wang et al., 2021).

6.5   Actions and benefits resulting from the implementation of a sustainable supply chain

An SSC is becoming increasingly important for companies in today's world and is thus a key feature of business strategy. It involves taking actions aimed at minimizing the negative impact of a company's activities on the natural environment, society, and the economy, while also striving to achieve operational and competitive efficiency. Implementing an SSC brings a range of benefits, including reducing a product's adverse environmental impact by limiting greenhouse gas emissions, optimizing the use of natural resources, and minimizing the amount of waste generated. Additionally, it helps improve a company's image, boost operational efficiency through better resource management, and build trust with customers and business partners. The long-term benefits of an SSC include increased competitiveness, operational stability, and a positive impact on the company's reputation on the market. Implementing such a system requires undertaking various steps at multiple stages of the logistics process. The examples of actions and benefits associated with the implementation of an SSC are presented in Table 6.2.

The benefits of implementing an SSC include limiting the negative impact on the natural environment, improving the company's reputation, lowering operational costs through more efficient resource utilization, and increasing trust among customers and business partners. As a result, companies can achieve a long-term sustainable competitive advantage.

6.6   Conclusions

In conclusion, based on an analysis of the literature, the authors propose a conceptual model of SSCM focused on designing sustainable products (Figure 6.2). The model incorporates performance criteria, which have been discussed in depth

TABLE 6.2 Activities and benefits of SSC implementation

Activities and benefits	Characteristics	References
Use of eco-friendly materials and resources	Transitioning to more environmentally friendly materials and resources helps reduce the negative impact on the natural environment. Benefits include reducing greenhouse gas emissions and natural resource consumption	Ashraf, Ahmed, et al. (2020), Ashraf, Saleem, et al. (2020) and Wang et al. (2020)
Transport optimization	Improving transport efficiency through optimized routes, using more eco-friendly modes of transportation (e.g., maritime transport instead of air transport), and reducing the number of empty trips help reduce CO_2 emissions and transport costs.	Patra (2018), Saada (2021), Sherif et al. (2020, 2021) and Zhao et al. (2020)
Increasing energy efficiency	Implementing measures to reduce energy consumption in production and logistics processes leads to a decrease in greenhouse gas emissions and operational costs	Centobelli et al. (2018), Huang et al. (2020), Mohtashami et al. (2020), Patra (2018) and Vegter et al. (2020, 2023)
Waste minimization and recycling	Reducing the amount of waste generated through sustainable production processes and promoting recycling and material reuse helps reduce environmental pollution.	Debnath and Sarkar (2023), Iqbal et al. (2020), Mastos et al. (2020) and Thomas and Mishra (2022)
Collaboration with local suppliers	Promoting collaboration with local suppliers results in lower CO_2 emissions associated with transportation and also supports local communities and economies.	Akhtar et al. (2023), Andalib Ardakani et al. (2023), Astanti et al. (2022) and Sharma et al. (2022)

Source: Authors' own table.

and which aim to incorporate sustainability throughout the entire supply chain. As a result, the model is made more practical and allows for a better assessment of sustainability within the supply chain. The SSCM model is based on the TBL concept. It consists of five blocks: (1) priorities in SSCM, (2) activities and benefits for SSC, (3) performance criteria for SSCM, (4) barriers for SSCM, and (5) a business approach to SSCM. The third block of SSCM incorporates performance criteria for SSCM along the entire supply chain, from designing sustainable products and ensuring sustainable ordering, purchasing, production processes, through collaboration with suppliers, and right up to reverse flow management.

One of the practical implications of this model is the implementation of stringent sustainability criteria at all stages of the supply chain process. Companies can utilize this model to assess and choose suppliers who adhere to sustainable

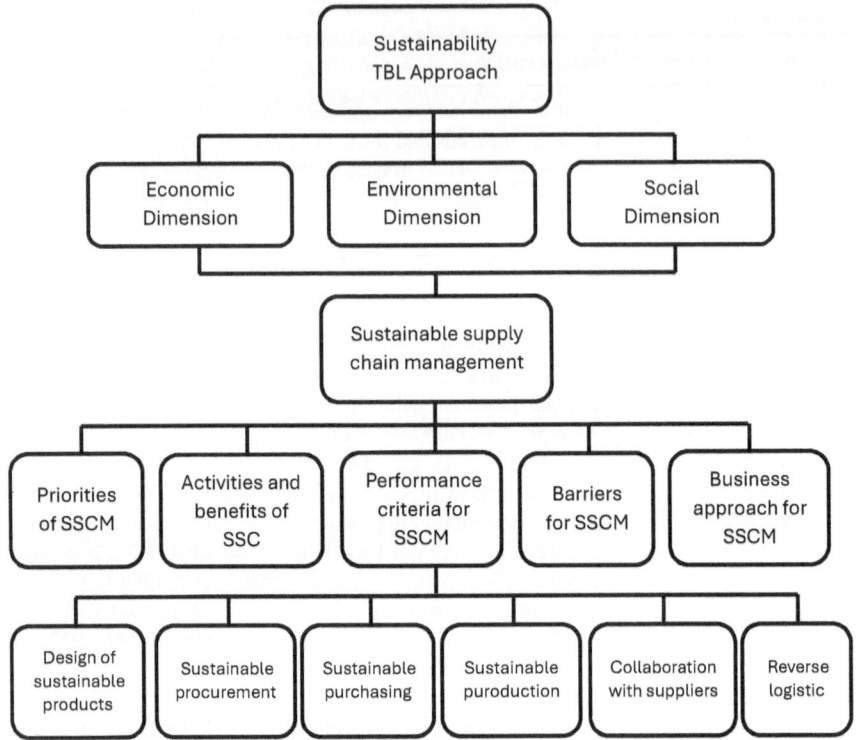

FIGURE 6.2   Conceptual model for SSCM.

Source: Own work.

production practices. Additionally, they can apply these criteria to product design, taking into account aspects of sustainable material sourcing, production, and waste management. By incorporating the performance criteria of an SSC, companies can bolster their commitment to sustainable development and help contain their negative environmental impact.

In summary, it should be noted that an SSC constitutes a crucial element in modern operations management. The concepts, structure, and essence of the SSC discussed in this chapter highlight its pivotal role in achieving business, social, and environmental goals. The definitions of sustainable logistics and reverse logistics, which are integral parts of the SSC, fit into the comprehensive version of this concept. The priorities of SSCM, including social responsibility, environmental protection, and economic efficiency, serve as a key reference point for actions in this area. The discussions on SP, with a particular emphasis on collaboration with suppliers in the design of sustainable products, demonstrate the importance of getting all participants in the supply chain involved in efforts to achieve sustainable development goals. Finally, the above-presented examples of the actions and benefits associated with an SSC underscore its potential for creating value for businesses and communities. An analysis of the issues connected with developing an SSC and identifying those factors exerting pressure on this area would ensure a comprehensive understanding of the challenges and opportunities associated with its effective management in the spirit of sustainable development.

In view of the above, future research directions should focus on the development of advanced methodologies for assessing the environmental impact of products, the implementation of digital technologies for monitoring and managing SSCs, as well as on the influence of climate and social change on strategies for designing sustainable products. Another important area of research would be new business models supporting sustainable products and the development of certification and standardization systems for sustainable products.

References

Abualigah, L., Hanandeh, E. S., Zitar, R. A., Thanh, C. Le, Khatir, S., & Gandomi, A. H. (2023). Revolutionizing sustainable supply chain management: A review of metaheuristics. *Engineering Applications of Artificial Intelligence, 126.* https://doi.org/10.1016/j.engappai.2023.106839

Adomako, S. (2020). Environmental collaboration, sustainable innovation, and small and medium-sized enterprise growth in sub-Saharan Africa: Evidence from Ghana. *Sustainable Development, 28*(6), 1609–1619.

Ageron, B., Gunasekaran, A., & Spalanzani, A. (2012). Sustainable supply management: An empirical study. *International Journal of Production Economics, 140*(1), 168–182. https://doi.org/10.1016/j.ijpe.2011.04.007

Ahi, P., & Searcy, C. (2013). A comparative literature analysis of definitions for green and sustainable supply chain management. *Journal of Cleaner Production, 52,* 329–341. https://doi.org/https://doi.org/10.1016/j.jclepro.2013.02.018

Akhtar, F., Huo, B., & Wang, Q. (2023). Embracing green supply chain collaboration through technologies: The bridging role of advanced manufacturing technology. *Journal of Business & Industrial Marketing, 38*(12), 2626–2642.

Andalib Ardakani, D., Soltanmohammadi, A., & Seuring, S. (2023). The impact of customer and supplier collaboration on green supply chain performance. *Benchmarking: An International Journal, 30*(7), 2248–2274.

Apeji, U. D., & Sunmola, F. T. (2022). Principles and factors influencing visibility in sustainable supply chains. *Procedia Computer Science, 200,* 1516–1527. https://doi.org/10.1016/j.procs.2022.01.353

Arora, A., Arora, A. S., Sivakumar, K., & Burke, G. (2020). Strategic sustainable purchasing, environmental collaboration, and organizational sustainability performance: The moderating role of supply base size. *Supply Chain Management. An International Journal, 25*(6), 709–728.

Ashraf, S., Ahmed, T., Saleem, S., & Aslam, Z. (2020). Diverging mysterious in green supply chain management. *Oriental Journal of Computer Science and Technology, 13*(1), 22–28.

Ashraf, S., Saleem, S., Chohan, A. H., Aslam, Z., & Raza, A. (2020). Challenging strategic trends in green supply chain management. *International Journal of Research in Engineering and Applied Sciences (JREAS), 5*(2), 71–74.

Astanti, R. D., Daryanto, Y., & Dewa, P. K. (2022). Low-carbon supply chain model under a vendor-managed inventory partnership and carbon cap-and-trade policy. *Journal of Open Innovation: Technology, Market, and Complexity, 8*(1), 30.

Azzi, R., Chamoun, R. K., & Sokhn, M. (2019). The power of a blockchain-based supply chain. *Computers & Industrial Engineering, 135*, 582–592.

Barbosa-Póvoa, A. P. (2014). Process supply chains management – where are we? Where to go next? *Frontiers in Energy Research, 2*(JUN). Frontiers Media S.A. https://doi.org/10.3389/fenrg.2014.00023

Barbosa-Póvoa, A. P., da Silva, C., & Carvalho, A. (2018). Opportunities and challenges in sustainable supply chain: An operations research perspective. *European Journal of Operational Research, 268*(2), 399–431. https://doi.org/10.1016/j.ejor.2017.10.036

Barrane, F. Z., Ndubisi, N. O., Kamble, S., Karuranga, G. E., & Poulin, D. (2021). Building trust in multi-stakeholder collaborations for new product development in the digital transformation era. *Benchmarking: An International Journal, 28*(1), 205–228.

Björklund, M., & Forslund, H. (2019). Challenges addressed by swedish third-party logistics providers conducting sustainable logistics business cases. *Sustainability, 11*(9), 2654.

Carter, C. R., & Rogers, D. S. (2008). A framework of sustainable supply chain management: Moving toward new theory. *International Journal of Physical Distribution & Logistics Management, 38*(5), 360–387. https://doi.org/10.1108/096000308 10882816

Centobelli, P., Cerchione, R., & Esposito, E. (2018). Environmental sustainability and energy-efficient supply chain management: A review of research trends and proposed guidelines. *Energies, 11*(2), 275.

Chen, L., Zhao, X., Tang, O., Price, L., Zhang, S., & Zhu, W. (2017). Supply chain collaboration for sustainability: A literature review and future research agenda. *International Journal of Production Economics, 194*, 73–87.

Das, S., & Hassan, H. M. K. (2022). Impact of sustainable supply chain management and customer relationship management on organizational performance. *International Journal of Productivity and Performance Management, 71*(6), 2140–2160. https://doi.org/10.1108/IJPPM-08-2020-0441

de Almeida Santos, D., Luiz Gonçalves Quelhas, O., Francisco Simões Gomes, C., Perez Zotes, L., Luiz Braga França, S., Vinagre Pinto de Souza, G., Amarante de Araújo, R., & da Silva Carvalho Santos, S. (2020). Proposal for a maturity model in sustainability in the supply chain. *Sustainability, 12*(22), 9655.

Debnath, A., & Sarkar, B. (2023). Effect of circular economy for waste nullification under a sustainable supply chain management. *Journal of Cleaner Production, 385*, 135477. https://doi.org/10.1016/j.jclepro.2022.135477

DeCampos, H. A., Fawcett, S. E., & Melnyk, S. A. (2022). Collaboration expectation gaps, transparency and integrated NPD performance: A multi-case study. *Journal of Purchasing and Supply Management, 28*(4), 100789.

Defra. (2006). *Procuring the Future–The Sustainable Procurement Task Force National Action Plan.* Department for Environment, Food and Rural Affairs, London.

Esfahbodi, A., Zhang, Y., Watson, G., & Zhang, T. (2017). Governance pressures and performance outcomes of sustainable supply chain management – An empirical analysis of UK manufacturing industry. *Journal of Cleaner Production, 155*, 66–78. https://doi.org/10.1016/j.jclepro.2016.07.098

Esmaeilian, B., Sarkis, J., Lewis, K., & Behdad, S. (2020). Blockchain for the future of sustainable supply chain management in Industry 4.0. *Resources, Conservation and Recycling, 163*, 105064. https://doi.org/10.1016/j.resconrec.2020.105064

Faramarzi-Oghani, S., Dolati Neghabadi, P., Talbi, E. G., & Tavakkoli-Moghaddam, R. (2023). Meta-heuristics for sustainable supply chain management: A review. *International Journal of Production Research*, *61*(6), 1979–2009. Taylor and Francis Ltd. https://doi.org/10.1080/00207543.2022.2045377

Fernando, Y., Halili, M., Tseng, M. L., Tseng, J. W., & Lim, M. K. (2022). Sustainable social supply chain practices and firm social performance: Framework and empirical evidence. *Sustainable Production and Consumption*, *32*, 160–172. https://doi.org/10.1016/j.spc.2022.04.020

Flores-Sigüenza, P., Marmolejo-Saucedo, J. A., Niembro-Garcia, J., & Lopez-Sanchez, V. M. (2021). A systematic literature review of quantitative models for sustainable supply chain management. *Mathematical Biosciences and Engineering*, *18*(3), 2206–2229.

Ganesh Kumar, R., & Ashlin Nimo, J. R. (2020). A conceptual framework for reverse logistics performance and innovation. *International Journal of Supply Chain Management*, *9*(1), 430.

Gao, J., Xiao, Z., Wei, H., & Zhou, G. (2020). Dual-channel green supply chain management with eco-label policy: A perspective of two types of green products. *Computers & Industrial Engineering*, *146*, 106613. https://doi.org/10.1016/j.cie.2020.106613

Genç, E., & Di Benedetto, C. A. (2015). Cross-functional integration in the sustainable new product development process. The role of the environmental specialist. *Industrial Marketing Management*, *50*, 150–161.

Ghadimi, P., Azadnia, A. H., Heavey, C., Dolgui, A., & Can, B. (2016). A review on the buyer–supplier dyad relationships in sustainable procurement context: Past, present and future. *International Journal of Production Research*, *54*(5), 1443–1462.

Gopal, P. R. C., & Thakkar, J. (2016). Sustainable supply chain practices: An empirical investigation on Indian automobile industry. *Production Planning & Control*, *27*(1), 49–64. https://doi.org/10.1080/09537287.2015.1060368

Govindan, K., Aditi, D. D. J., Kaul, A., & Jha, P. C. (2021). Structural model for analysis of key performance indicators for sustainable manufacturer–supplier collaboration: A grey-decision-making trial and evaluation laboratory-based approach. *Business Strategy and the Environment*, *30*(4), 1702–1722.

Govindan, K., Mina, H., Esmaeili, A., & Gholami-Zanjani, S. M. (2020). An integrated hybrid approach for circular supplier selection and closed loop supply chain network design under uncertainty. *Journal of Cleaner Production*, *242*. https://doi.org/10.1016/j.jclepro.2019.118317

Haessler, P. (2020). Strategic decisions between short-term profit and sustainability. *Administrative Sciences*, *10*(3). https://doi.org/10.3390/admsci10030063

Heidary Dahooie, J., Zamani Babgohari, A., Meidutė-Kavaliauskienė, I., & Govindan, K. (2021). Prioritising sustainable supply chain management practices by their impact on multiple interacting barriers. *International Journal of Sustainable Development and World Ecology*, *28*(3), 267–290. https://doi.org/10.1080/13504509.2020.1795004

Houé, T., & Duchamp, D. (2021). Relational impact of buyer–supplier dyads on sustainable purchasing and supply management: A proximity perspective. *The International Journal of Logistics Management*, *32*(2), 567–591.

Huang, Y.-S., Fang, C.-C., & Lin, Y.-A. (2020). Inventory management in supply chains with consideration of logistics, green investment and different carbon emissions policies. *Computers & Industrial Engineering*, *139*, 106207. https://doi.org/10.1016/j.cie.2019.106207

Iqbal, M. W., Kang, Y., & Jeon, H. W. (2020). Zero waste strategy for green supply chain management with minimization of energy consumption. *Journal of Cleaner Production*, *245*, 118827.

Islam, M. H., Sarker, M. R., Hossain, M. I., Ali, K., & Noor, K. M. A. (2020). Towards Sustainable Supply Chain Management (SSCM): A case of leather industry. *Journal of Operations and Strategic Planning*, *3*(1), 81–98. https://doi.org/10.1177/2516600x 20924313

Islam, M. M., Turki, A., Murad, M. W., & Karim, A. (2017). Do sustainable procurement practices improve organizational performance? *Sustainability (Switzerland)*, *9*(12). https://doi.org/10.3390/su9122281

Kannan, D. (2021). Sustainable procurement drivers for extended multi-tier context: A multi-theoretical perspective in the Danish supply chain. *Transportation Research Part E: Logistics and Transportation Review*, *146*, 102092. https://doi.org/10.1016/j. tre.2020.102092

Kaur, H., & Singh, S. P. (2019). Sustainable procurement and logistics for disaster resilient supply chain. *Annals of Operations Research*, *283*, 309–354.

Khan, S. A. R., Yu, Z., Golpira, H., Sharif, A., & Mardani, A. (2021). A state-of-the-art review and meta-analysis on sustainable supply chain management: Future research directions. *Journal of Cleaner Production*, *278*, 123357. https://doi.org/10.1016/J. JCLEPRO.2020.123357

Kot, S. (2018). Sustainable supply chain management in small and medium enterprises. *Sustainability (Switzerland)*, *10*(4). https://doi.org/10.3390/su10041143

Krishnan, R., Agarwal, R., Bajada, C., & Arshinder, K. (2020). Redesigning a food supply chain for environmental sustainability – An analysis of resource use and recovery. *Journal of Cleaner Production*, *242*, 118374. https://doi.org/10.1016/j.jclepro.2019.118374

Krzysztofek, A. (2014). Zrównoważone zarządzanie łańcuchem dostaw jako element wdrażania społecznej odpowiedzialności. *Logistyka*, *5*, 1939–1949.

Kshetri, N. (2021). Blockchain and sustainable supply chain management in developing countries. *International Journal of Information Management*, *60*. https://doi.org/10.1016/ j.ijinfomgt.2021.102376

Leal Filho, W., Viera Trevisan, L., Paulino Pires Eustachio, J. H., Dibbern, T., Castillo Apraiz, J., Rampasso, I., Anholon, R., Gornati, B., Morello, M., & Lambrechts, W. (2023). Sustainable supply chain management and the UN sustainable development goals: exploring synergies towards sustainable development. *The TQM Journal*, *Ahead-of-Print* (ahead-of-print). https://doi.org/10.1108/TQM-04-2023-0114

Letunovska, N., Offei, F. A., Junior, P. A., Lyulyov, O., Pimonenko, T., & Kwilinski, A. (2023). Green supply chain management: The effect of procurement sustainability on reverse logistics. *Logistics*, *7*(3), 47.

Lis, A., Sudolska, A., & Tomanek, M. (2020). Mapping research on sustainable supply-chain management. *Sustainability (Switzerland)*, *12*(10). https://doi.org/10.3390/SU12103987

Liu, L., Song, W., & Liu, Y. (2023). Leveraging digital capabilities toward a circular economy: Reinforcing sustainable supply chain management with Industry 4.0 technologies. *Computers and Industrial Engineering*, *178*. https://doi.org/10.1016/j.cie.2023.109113

Luthra, S., & Mangla, S. K. (2018). Evaluating challenges to Industry 4.0 initiatives for supply chain sustainability in emerging economies. *Process Safety and Environmental Protection*, *117*, 168–179.

Mackelprang, A. W., Bernardes, E., Burke, G. J., & Welter, C. (2018). Supplier innovation strategy and performance: A matter of supply chain market positioning. *Decision Sciences*, *49*(4), 660–689.

Mageto, J. (2021). Big data analytics in sustainable supply chain management: A focus on manufacturing supply chains. *Sustainability (Switzerland)*, *13*(13). MDPI. https://doi. org/10.3390/su13137101

Mahadevan, K. (2019). Collaboration in reverse: A conceptual framework for reverse logistics operations. *International Journal of Productivity and Performance Management*, *68*(2), 482–504.

Mastos, T. D., Nizamis, A., Vafeiadis, T., Alexopoulos, N., Ntinas, C., Gkortzis, D., Papadopoulos, A., Ioannidis, D., & Tzovaras, D. (2020). Industry 4.0 sustainable supply chains: An application of an IoT enabled scrap metal management solution. *Journal of Cleaner Production*, *269*, 122377. https://doi.org/10.1016/j.jclepro.2020.122377

Mehdikhani, R., & Valmohammadi, C. (2019). Strategic collaboration and sustainable supply chain management: The mediating role of internal and external knowledge sharing. *Journal of Enterprise Information Management*, *32*(5), 778–806.

Melander, L. (2018). Customer and supplier collaboration in green product innovation: External and internal capabilities. *Business Strategy and the Environment*, *27*(6), 677–693.

Men, F., Yaqub, R. M. S., Yan, R., Irfan, M., & Haider, A. (2023). The impact of top management support, perceived justice, supplier management, and sustainable supply chain management on moderating the role of supply chain agility. *Frontiers in Environmental Science*, *10*. https://doi.org/10.3389/fenvs.2022.1006029

Messah, Y., Wirahadikusumah, R., & Abduh, M. (2023). Structural equation model (SEM) of the factors affecting sustainable procurement for construction work. *International Journal of Construction Management*, *23*(13), 2221–2229.

Messah, Y. A., Abduh, M., & Wirahadikusumah, R. D. (2023). Conceptual framework for sustainable procurement of construction works. *International Journal of Procurement Management*, *17*(4), 488–506.

Miemczyk, J., Johnsen, T. E., & Macquet, M. (2012). Sustainable purchasing and supply management: A structured literature review of definitions and measures at the dyad, chain and network levels. *Supply Chain Management: An International Journal*, *17*(5), 478–496.

Mohtashami, Z., Aghsami, A., & Jolai, F. (2020). A green closed loop supply chain design using queuing system for reducing environmental impact and energy consumption. *Journal of Cleaner Production*, *242*, 118452.

Mughal, Y. H., Nair, K. S., Arif, M., Albejaidi, F., Thurasamy, R., Chuadhry, M. A., & Malik, S. Y. (2023). Employees' perceptions of green supply-chain management, corporate social responsibility, and sustainability in organizations: Mediating effect of reflective moral attentiveness. *Sustainability (Switzerland)*, *15*(13). https://doi.org/10.3390/su151310528

Murali, S., Balasubramanian, M., & Choudary, M. V. (2023). Investigation on the impact of the supplier, customer, and organization collaboration factors on the performance of new product development. *International Journal of System Assurance Engineering and Management*, *14*(Suppl 4), 918–923.

Muthugala, S., & Nayagam, N. (2012). Incorporating the triple bottom line in public procurement. In *Balancing social, environmental and economic considerations in procurement*. United Nations, New York, 30–33.

Niu, B., & Mu, Z. (2020). Sustainable efforts, procurement outsourcing, and channel co-opetition in emerging markets. *Transportation Research Part E: Logistics and Transportation Review*, *138*. https://doi.org/10.1016/j.tre.2020.101960

Oliveira, G. A., Tan, K. H., & Guedes, B. T. (2018). Lean and green approach: An evaluation tool for new product development focused on small and medium enterprises. *International Journal of Production Economics*, *205*, 62–73.

Parker, D. B., Zsidisin, G. A., & Ragatz, G. L. (2008). Timing and extent of supplier integration in new product development: A contingency approach. *Journal of Supply Chain Management*, *44*(1), 71–83.

Patra, P. K. (2018). Green logistics: Eco-friendly measure in supply-chain. *Management Insight – The Journal of Incisive Analysers, 14*(1), 65–71.

Paul, A., Shukla, N., Paul, S. K., & Trianni, A. (2021). Sustainable supply chain management and multi-criteria decision-making methods: A systematic review. *Sustainability (Switzerland), 13*(13). MDPI. https://doi.org/10.3390/su13137104

Persdotter Isaksson, M., Hulthén, H., & Forslund, H. (2019). Environmentally sustainable logistics performance management process integration between Buyers and 3PLs. *Sustainability, 11*(11), 3061.

Petersen, K. J., Handfield, R. B., & Ragatz, G. L. (2005). Supplier integration into new product development: Coordinating product, process and supply chain design. *Journal of Operations Management, 23*(3–4), 371–388.

Preuss, L. (2009). Addressing sustainable development through public procurement: The case of local government. *Supply Chain Management: An International Journal, 14*(3), 213–223. https://doi.org/10.1108/13598540910954557

Prier, E., Schwerin, E., & McCue, C. P. (2016). Implementation of sustainable public procurement practices and policies: A sorting framework. *Journal of Public Procurement, 16*(3), 312–346. https://doi.org/10.1108/JOPP-16-03-2016-B004

Saada, R. (2021). *Green transportation in green supply chain management.* Green Supply Chain-Competitiveness and Sustainability, IntechOpen, London, 25–44.

Sajjad, A., Eweje, G., & Tappin, D. (2020). Managerial perspectives on drivers for and barriers to sustainable supply chain management implementation: Evidence from New Zealand. *Business Strategy and the Environment, 29*(2), 592–604.

Sánchez-Flores, R. B., Cruz-Sotelo, S. E., Ojeda-Benitez, S., & Ramírez-Barreto, M. E. (2020). Sustainable supply chain management—A literature review on emerging economies. *Sustainability (Switzerland), 12*(17). MDPI. https://doi.org/10.3390/SU12176972

Schoenherr, T., & Wagner, S. M. (2016). Supplier involvement in the fuzzy front end of new product development: An investigation of homophily, benevolence and market turbulence. *International Journal of Production Economics, 180*, 101–113.

Schulze, H., & Bals, L. (2020). Implementing sustainable purchasing and supply management (SPSM): A Delphi study on competences needed by purchasing and supply management (PSM) professionals. *Journal of Purchasing and Supply Management, 26*(4), 100625. https://doi.org/10.1016/j.pursup.2020.100625

Sembiring, N., Tambunan, M. M., & Ginting, E. (2020). Analysing company's performance by using sustainable supply chain management (SSCM). *IOP Conference Series: Materials Science and Engineering, 852*(1). https://doi.org/10.1088/1757-899X/852/1/012108

Seuring, S., & Müller, M. (2008). From a literature review to a conceptual framework for sustainable supply chain management. *Journal of Cleaner Production, 16*(15), 1699–1710. https://doi.org/10.1016/j.jclepro.2008.04.020

Seuring, S., Sarkis, J., Müller, M., & Rao, P. (2008). Sustainability and supply chain management – An introduction to the special issue. *Journal of Cleaner Production, 16*(15), 1545–1551 https://doi.org/10.1016/j.jclepro.2008.02.002

Sharma, M., Luthra, S., Joshi, S., & Kumar, A. (2022). Developing a framework for enhancing survivability of sustainable supply chains during and post-COVID-19 pandemic. *International Journal of Logistics Research and Applications, 25*(4–5), 433–453.

Shekarian, E., Ijadi, B., Zare, A., & Majava, J. (2022). Sustainable supply chain management: A comprehensive systematic review of industrial practices. *Sustainability, 14*(13), 7892.

Sherif, S. U., Asokan, P., Sasikumar, P., Mathiyazhagan, K., & Jerald, J. (2021). Integrated optimization of transportation, inventory and vehicle routing with simultaneous pickup and delivery in two-echelon green supply chain network. *Journal of Cleaner Production, 287*, 125434.

Sherif, S. U., Sasikumar, P., Asokan, P., & Jerald, J. (2020). An eco-friendly closed loop supply chain network with multi-facility allocated centralized depots for bidirectional flow in a battery manufacturing industry. *Journal of Advances in Management Research, 17*(1), 131–159.

Silva, M. E., & Figueiredo, M. D. (2020). Practicing sustainability for responsible business in supply chains. *Journal of Cleaner Production, 251*, 119621. https://doi.org/https://doi.org/10.1016/j.jclepro.2019.119621

Slot, J. H., Wuyts, S., & Geyskens, I. (2020). Buyer participation in outsourced new product development projects: The role of relationship multiplexity. *Journal of Operations Management, 66*(5), 578–612.

Sroufe, R., Curkovic, S., Montabon, F., & Melnyk, S. A. (2000). The new product design process and design for environment: "Crossing the chasm." *International Journal of Operations & Production Management, 20*(2), 267–291.

Straka, M. (2019). *Distribution and supply logistics*. Cambridge Scholars Publishing, Newcastle upon Tyne, 609.

Teixeira, C. R. B., Assumpção, A. L., Correa, A. L., Savi, A. F., & Prates, G. A. (2018). The contribution of green logistics and sustainable purchasing for green supply chain management. *Independent Journal of Management & Production, 9*(3), 1002–1026.

Thomas, A., & Mishra, U. (2022). A sustainable circular economic supply chain system with waste minimization using 3D printing and emissions reduction in plastic reforming industry. *Journal of Cleaner Production, 345*, 131128. https://doi.org/10.1016/j.jclepro.2022.131128

Thomson, J., & Jackson, T. (2007). Sustainable procurement in practice: Lessons from local government. *Journal of Environmental Planning and Management, 50*(3), 421–444. https://doi.org/10.1080/09640560701261695

Tsai, F. M., Bui, T. D., Tseng, M. L., Ali, M. H., Lim, M. K., & Chiu, A. S. (2021). Sustainable supply chain management trends in world regions: A data-driven analysis. *Resources, Conservation and Recycling, 167*. https://doi.org/10.1016/j.resconrec.2021.105421

Vegter, D., van Hillegersberg, J., & Olthaar, M. (2020). Supply chains in circular business models: Processes and performance objectives. *Resources, Conservation and Recycling, 162*, 105046.

Vegter, D., van Hillegersberg, J., & Olthaar, M. (2023). Performance measurement system for circular supply chain management. *Sustainable Production and Consumption, 36*, 171–183. https://doi.org/10.1016/j.spc.2023.01.003

Walker, H., & Brammer, S. (2009). Sustainable procurement in the United Kingdom public sector. *Supply Chain Management: An International Journal, 14*(2), 128–137. https://doi.org/10.1108/13598540910941993

Wang, B., Luo, W., Zhang, A., Tian, Z., & Li, Z. (2020). Blockchain-enabled circular supply chain management: A system architecture for fast fashion. *Computers in Industry, 123*, 103324.

Wang, D.-F., Dong, Q.-L., Peng, Z.-M., Khan, S. A. R., & Tarasov, A. (2018). The green logistics impact on international trade: Evidence from developed and developing countries. *Sustainability, 10*(7), 2235.

Wang, Y., Modi, S. B., & Schoenherr, T. (2021). Leveraging sustainable design practices through supplier involvement in new product development: The role of the suppliers' environmental management capability. *International Journal of Production Economics*, *232*, 107919. https://doi.org/10.1016/j.ijpe.2020.107919

Wynstra, F., Anderson, J. C., Narus, J. A., & Wouters, M. (2012). Supplier development responsibility and NPD project outcomes: The roles of monetary quantification of differences and supporting-detail gathering. *Journal of Product Innovation Management*, *29*, 103–123.

Yang, Y., & Wang, Y. (2020). Supplier selection for the adoption of green innovation in sustainable supply chain management practices: A case of the Chinese textile manufacturing industry. *Processes*, *8*(6), 1–24. https://doi.org/10.3390/pr8060717

Zhang, D., Frei, R., Wills, G., Gerding, E., Bayer, S., & Senyo, P. K. (2023). Strategies and practices to reduce the ecological impact of product returns: An environmental sustainability framework for multichannel retail. *Business Strategy and the Environment*, *32*(7), 4636–4661. https://doi.org/10.1002/bse.3385

Zhao, X., Ke, Y., Zuo, J., Xiong, W., & Wu, P. (2020). Evaluation of sustainable transport research in 2000–2019. *Journal of Cleaner Production*, *256*, 120404.

7

DIGITALISATION IN SUSTAINABLE PRODUCTS AND SUSTAINABLE PRODUCTION

Beata Paliwoda and Sergiusz Strykowski

7.1   Sustainable products

The transformation of products from conventional to sustainable, along with the general classification of product types, is detailed in the second chapter of this monograph. The general classification includes conventional products, bioproducts, ecologic products, and sustainable products. For many years, the objective has been to incorporate sustainable approaches into products and manufacturing processes to reduce emissions and achieve a harmonious equilibrium among environmental, social, and economic aspects of the triple bottom line. As presented in Chapter 2, a traditional approach to sustainable products is aimed to minimise their impact while fulfilling present requirements without jeopardising the needs of future generations. Sustainability is an equilibrium – a sweet spot where negative impacts are not just minimised but actively offset. However, in the nowadays context, simply minimising the negative impacts is no longer sufficient. In more forward-thinking model, sustainable products must evolve into restorative and regenerative products, which represent a "higher level of sustainability". Such products not only have a positive impact on society and environment but also actively give back, eventually tipping the scale towards net positive outcomes (Sphera, 2024).

The need for adopting sustainable practices pushes companies to innovate and create new systems and procedures. This practice not only contributes to the mitigation of environmental effects but also efficiently fulfils consumer demands, hence promoting market competition. In this regard, the restorative and regenerative products are the most preferable in terms of eco-friendliness. Not only do they avoid harming the environment, but they also generate a positive impact. This influence outweighs any negative effects from the production processes and

DOI: 10.4324/9781032710693-8

products contribute to improving the condition of the environment while balancing economic and ecological considerations.

7.1.1   Ways of achieving sustainable products

Companies seeking a sustainable future can adopt environmentally conscious design principles. This involves integrating environmental considerations into the design and development process, to minimise adverse environmental impact throughout the entire product life cycle. Life cycle assessment (LCA) is a cradle-to-grave analysis technique to assess environmental impacts associated with all the stages of a product's life (Muralikrishna & Manickam, 2017). Incorporating this approach is essential in creating sustainable products (Table 7.1).

Another widely used approach that organisations take is transitioning towards a circular economy. A circular economy is an economic form with material circular flow as its core, and it attempts to reduce the influences of human activity on the natural environment system via the flow of closed material in the human socio-economic system (Mao et al., 2018). The circular economy concept was proposed by the European Union and China to provide future generations with healthier and sustainable ecosystems by closing the loop of the product lifecycle (Cifuentes-Faura, 2022; Zhu et al., 2022). The European Union has put into action a plan regarding plastic within the circular economy, which has introduced the concept of Extended Producer Responsibility where producers must bear both financial and organisational accountability for managing their products when they reach the end of their life cycle – including tasks such as collection, sorting, and treatment (Lorang et al., 2022). Unlike sustainable production, which aims to minimise landfill waste by conserving raw materials and resources, the circular economy operates on the premise of efficiently utilising natural resources throughout a product's lifespan, ultimately aiming for zero waste (de Angelis, Howard & Miemczyk, 2018). The principles of the circular economy focus on designing products and processes that minimise waste and make the most of resources. These principles are often depicted in a hierarchy known as the "waste hierarchy" or sometimes the "3R hierarchy", consisting of three main stages: reduce, reuse, and recycle. However, those three main stages can be further expanded by adding additional steps and explanations for each stage (the "9R", see Table 7.2).

Transitioning to a circular economy may be challenging for manufacturers and producers, as this includes navigating through various design options and assessing the actual environmental benefits. Quantitative assessments, like LCA, can help by evaluating different strategies and identifying potential trade-offs (Ingemarsdotter & Dumont, 2022). LCA based on the ISO 14040 standard (ISO 14040:2009) is one of the most effective techniques to measure the circularity and environmental efficiency on a micro level. This methodology is widely used to assess the environmental performance of products, services, and systems and to compare different processes according to their environmental impacts, especially in the context

TABLE 7.1 Examples of improvements in design processes using a lifecycle approach

Design for material sourcing	Design for transport and distribution	Design for manufacture	Design for Product Use	Design for end-of-life
• Reduce weight • Increase reuse of products, components and sub-assemblies • Increase use of recycled materials • Reduce the use of scarce materials • Minimise/eliminate the use of substances hazardous to health or the environment • Decrease the need for consumables • Decrease the quantity of energy used throughout the product's life cycle (e.g., lighting) • Specify materials that emit low or zero volatile organic compounds (VOCs) • Use materials with a low environmental footprint	• Minimise product size and weight • Optimise shape and volume for maximum packing density • Optimise transport in relations to energy efficiency and emissions • Maximise reuse of packaging where possible • Reduce embodied energy in packaging • Use packaging that emits low or zero VOCs • Increase use of recycled materials in packaging • Increase the sharing rate (ride share options) of commuting cars	• Reduce energy consumption • Reduce consumption of natural resources • Reduce process waste • Use internally recovered or recycled materials from process waste • Reduce emissions to air water and soil during manufacture • Consider reducing the number of parts • Reduce use of hazardous process chemicals (e.g., volatile solvents)	• Reduce energy consumption in use (e.g., lighting) • Reduce consumption of natural resources in use (e.g., water in use) • Optimise quantity and nature of consumables • Maximise product lifetime by designing for durability and reliability • Maximise product lifetime by designing for ease of maintenance and reparability • Maximise product lifetime by designing for refurbishment/remanufacturing • Minimise/eliminate hazardous substances during use	• Restrict use of substances classified as hazardous • Maximise the ability to reuse and recycle components and materials by design for disassembly • Minimise design aspects detrimental to reuse and recycling (e.g., mixture of materials) • Reduce amount of residual waste generated • Reduce energy and water required for disassembly and recycling

Source: Based on ISO 14006:2020 and IEC 62430:2019.

TABLE 7.2 The different stages that R-Strategies can be implemented

Area	Title	Description
Smarter design, supply chains and manufacturing	R0 Refuse	Make a product redundant by abandoning its function or by offering the same function by a radically different (e.g., digital) product or service
	R1 Rethink	Make product use more intensive (e.g., through product-as-a-service, reuse and sharing models or putting multi-functional products on the market)
	R2 Reduce	Increase efficiency in product manufacture or use by consuming fewer natural resources and materials
Extend the lifespan of products and their parts	R3 Reuse	Reuse of a product which is still in good condition and fulfils its original function (and is not waste) for the same purpose for which it was conceived
	R4 Repair	Repair and maintenance of defective products so they can be used with their original function
	R5 Refurbish	Restore an old product and bring it up to date (to a specified quality level)
	R6 Remanufacture	Use parts of a discarded product in a new product with the same function (as-new-condition)
	R7 Repurpose	Use a redundant product or its parts in a new product with a different function.
Useful application of materials	R8 Recycle	Process materials from waste into new products, materials or substances whether for the original or other purposes
	R9 Recover	Incineration of materials with energy recovery or reprocessing into materials that are to be used as fuels

Source: PBL Netherlands Environmental Assessment Agency (2017).

of product end of life (EoL; Iacovidou et al., 2017). While LCA was originally conceived as only applicable to products, its benefits and potential can also be extended to the assessment of organisations. An organisational LCA, based on technical specification ISO/TS 14072 (ISO/TS14072:2014), can provide insights into the value chain and identify hotspots where action is required (UNEP, Life Cycle Initiative, 2015). LCA is widely used to assess material circularity.

While the creation of sustainable and restorative products brings numerous benefits, there remain significant challenges that companies may encounter. The high price and limited availability of sustainable materials is one of them.

Procuring these materials often involves higher expenses compared to conventional counterparts, impacting the overall production costs and potentially leading to higher prices for consumers. Another challenge is the need for innovative design and manufacturing processes to meet sustainability goals. This would include relocating resources towards investments in new technologies and restructuring of the production lines. Another considerable obstacle in making sustainable products is a life-cycle approach. This means designing products to last longer, responsibly handling their disposal at the end of their life and evaluating the environmental impact at every step. Accomplishing this requires data collection, infrastructure investments, and cooperation among different industries. And last, but not least, while the circular economy aims to reduce waste and promote sustainability, there may be significant health risks associated with the recycling processes, particularly due to the unknown chemical compositions of a product that was recycled multiple times and their long-term effects on human health, for example, the research on recycled tyre playgrounds found to have potential toxic and carcinogenic impacts on children's health (Llompart et al., 2013; Winz et al., 2023; Lopez-Galvez et al., 2022). This can lead to a product being sustainable from the ecological point of view, but unsustainable from the social point of view.

7.2   Digitalisation in sustainable products

Digitalisation, as defined, refers to the adaptation of a system, process, etc., to be operated with the use of computers, and the Internet. Considering the Fourth Industrial Revolution (Industry 4.0), which focuses on developing technology and implementing automation into production lines, embracing digitalisation becomes even more crucial. The Internet of Things (IoT) is one of the key technologies of the Industry 4.0, which is focused on real-time data monitoring. Digitalisation in production, through the incorporation of IoT solutions, leads to a better understanding of processes, contributing to efficient and sustainable manufacturing by minimising scraps, losses, and wastes.

Digitalisation in sustainable product development represents a transformative approach, leveraging digital technologies to create products that are not only innovative but also environmentally friendly and socially responsible. This process involves integrating digital tools, data analytics, and advanced technologies such as artificial intelligence (AI) and the Industrial Internet of Things (IIoT) to optimise product design, manufacturing, and lifecycle management with sustainability at the core.

One of the key benefits of digitisation is the enhanced ability to track and analyse the entire lifecycle of products. By using sensors and IoT technology, companies can monitor how products are used and how they impact the environment throughout their lifespan. This data can then inform the designers of future products, making them more efficient and less harmful to the environment.

The conceptual model illustrating the application of digitalisation for achieving sustainable production and product usage is presented in Figure 7.1. In the model,

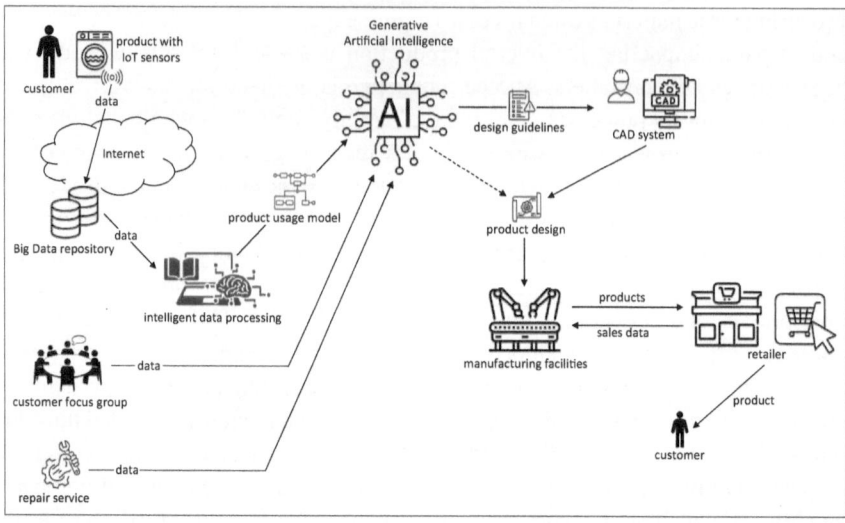

FIGURE 7.1 The conceptual model illustrating the application of digitalisation for achieving sustainable production and product usage.

the concept of production and product usage is broadly understood and encompasses the following phases:

- Monitoring of product usage
- Analysis of data collected during the monitoring
- Generating guidelines for sustainable-aware changes and improvements in the product design
- Product design
- Physical production
- Distribution and sales

Those phases will be discussed in subsequent sections.

7.2.1 Product usage monitoring

Monitoring product usage is enabled through the application of IoT sensors connected to dedicated product elements. Monitoring includes collecting real-time data that describes details about each use of the product, such as date, time, functions, and options selected. For instance, with a washing machine, this would include settings for each laundry cycle – water temperature, spin speed, washing plan, load weight, and the amount of detergent and additional substances applied, such as bleach or fabric softener. All collected data is transmitted wirelessly over the Internet and stored in databases on the manufacturer's servers.

Additionally, customers are encouraged, by offering additional benefits, to register their purchases – registration naturally includes providing personal data, such as age, place of residence, and education. In this way, the manufacturer possesses both data on how the product is used and data on the demographic and social profile of the customer.

7.2.2   Data analysis

Since the amount of collected in IoT and transferred to the Cloud data is significant, Big Data technology must be applied and data management solutions such as Big Data analytics (BDA) must be used. Big Data refers to vast volumes of structured and unstructured data that are too large and complex to be stored and processed using traditional databases and data processing methods.

The collected data is then subjected to comprehensive examination using intelligent data analysis methods to extract meaningful patterns, trends, and correlations. These methods include statistical analysis, data mining, deep learning algorithms (neural network algorithms), and process mining. As a result, a digital model of product usage is created.

The second source of data on product usage is marketing research, such as surveys, focus groups, or interviews with customers. This source mainly allows for the collection of qualitative data (so-called soft data), such as the feelings and emotions of customers associated with using products – these data are often referred to as customer experiences during product usage. At this stage, the AI can be used for the so-called sentiment analysis. It is, of course, also possible to collect quantitative data, but it will not be as accurate as data from IoT sensors.

The third important source of data is information from repair services. This source provides data on defects and malfunctions that have emerged during product usage.

7.2.3   Generative artificial intelligence

The data gathered from these three sources form the input stream for generative AI. Generative AI refers to a class of AI models and algorithms designed to generate new content, such as images, text, audio, or even complex structures, that mimic human-created content.

Using the data received, generative AI can generate guidelines for changes and improvements to the product design and their functions, making them more environmentally friendly for daily use by reducing their ecological footprint. The examples of such changes are as follows:

• Energy efficiency improvements: products, especially electronic devices and appliances, can be designed to consume less energy during use.

- Use of recycled materials: products can be made from recycled materials, reducing the demand for virgin resources.
- Durability and reparability: products can be designed to last longer and be easily repairable reducing waste and the need for frequent replacements.
- Biodegradable materials: using materials that can break down naturally in the environment for products and packaging can reduce pollution. Examples include biodegradable plastics made from plant-based materials as described in Chapter 2.
- Reduced packaging: minimising the amount of packaging used for products or using recyclable or compostable packaging materials can significantly reduce waste.
- Water efficiency improvements: for appliances like washing machines and dishwashers, improving water efficiency to use less water per cycle without compromising performance.
- Eco-friendlier chemicals: avoiding harmful chemicals in products and opting for natural or less harmful alternatives can reduce environmental and health risks. For instance, using non-toxic, water-based paints and finishes.
- Smart technology: incorporating smart technology into products to optimise their energy usage. For example, smart thermostats that adjust heating and cooling based on occupancy and weather.
- EoL management: designing products with a plan for their EoL, such as easy disassembly for recycling or composting.

Since the guidelines for changes and improvements are based on data describing the actual daily use of products in real conditions, their quality will be significantly higher than those based solely on customer surveys and focus studies. Additionally, generative AI can also be used to conduct a product risk analysis aimed at reducing health risks associated with the recycling processes. This is particularly important due to the unknown chemical compositions of products that have been recycled multiple times and their long-term effects on human health.

7.2.4   Product design

The guidelines are then used to design the new version of a product. The design is developed entirely digitally using the capabilities of computer-aided design (CAD) systems. These advanced digital systems not only enable the development of the design but also allow for conducting various types of simulations and operational tests. Leveraging the CAD system eliminates the need for constructing physical prototypes. This not only streamlines the development process but also significantly reduces the environmental burden and impact associated with the production of prototypes. Through CAD systems, engineers can model various iterations, simulating airflow dynamics to optimise performance. In parallel, generative AI algorithms can explore countless design possibilities, factoring in variables such as weight distribution, structural integrity, optimise space utilisation, energy efficiency, and aesthetic appeal ultimately proposing novel solutions. Additionally, the use of virtual reality systems, such as the Oculus, allows engineers and designers to immerse themselves in a 3D environment where they can interact with and refine the digital prototype in real-time. This immersive experience further improves the ability to identify potential issues, make adjustments, and ensure the design meets all specifications before moving to production.

In a more advanced version, the need for human involvement and intervention in product design using CAD systems can be entirely omitted and the design can be autonomously generated by generative AI. This technology has the potential to autonomously generate designs based on specified parameters and constraints. As of 2024, these applications are still in the stages of experimentation or preliminary prototyping stages. However, the rapid advancements and the trajectory of generative AI's development suggest that the widespread implementation of such technologies is inevitable.

7.2.5 Manufacturing

The product design is then transmitted to production lines that function largely autonomously, thanks to computer-aided manufacturing (CAM) systems. These are digitally controlled manufacturing machines, robots, and industrial automation collectively known as Industry 4.0 solutions. Within such manufacturing setups, the IIoT leads to the creation of Smart Factories. Here, various objects such as machinery, devices, and products are embedded with sensors interconnected with each other and with the Internet. Through data collection, analysis, and exchange, machine-to-machine (M2M) learning occurs, allowing for adjustments in modes and settings without the need for human intervention. The IIoT extensively utilise sensors and actuators including position sensors, motion sensors, biosensors, mass or volume sensors, measurement sensors, and environment sensors. Also, the application of radio frequency identification solutions, which automatically identifies objects using radio waves, helps in warehouse management, inventory management, tool management, supply chain management, process monitoring and control, and life cycle management. A tag transmits its identity to a reader, either attached to a product or embedded within it. The tag carries a unique code containing information. As tagged items move through a reader, data is collected and sent to a database. In manufacturing, companies also use IT systems and dedicated software to create processes and manage quality management system documentation and dedicated software enabling digital operating instructions, and electronic documents on the workstations and paperless shop floor. The objective of a paperless factory is to eliminate paper-based processes and digitise the entire manufacturing process. Within a paperless workshop, all production-related information, including work orders, standard operating procedures, and quality control data, is stored electronically. The removal of paper-based methods improves interdepartmental communication and collaboration, reduce errors, increase productivity, save time, and improve product quality.

The integration of CAM in production lines enhance decision-making, improve performance, productivity, efficiency, quality and precision, reducing waste, and energy consumption. It allows for monitoring raw materials, work-in-progress, and final products, reduces human errors, and decreases equipment maintenance time and costs. Industry 4.0 solutions like Smart Factories, facilitated by the IIoT, optimise resource usage through real-time monitoring and automated adjustments. With sensors embedded in machinery and products, coupled with M2M learning sustainable production practices are promoted by enabling proactive maintenance of equipment and tooling, process optimisation, and minimising environmental impact by minimising scraps, losses, and wastes.

7.2.6   Distribution and sales

The next phase is distribution and sales. If the manufacturer collaborates closely with retailers, it gains online access to data from their sales management systems, especially data that reveals daily sales levels. This information allows for the precise determination of production levels, favouring short production series in a just-in-time distribution system while reducing long production runs meant for stockpiling and long-term storage of finished goods. Short production runs and just-in-time distribution inherently support sustainable development. However, implementing these solutions requires tight cooperation with retailers and the exchange of data on daily sales levels. Given that a specific manufacturer often works with several retailers who compete with each other, such data sharing in business practice is rare due to limited trust among parties and is referred to as a "romantic vision of full openness between partners in the supply chain".

Retailers conduct sales to end customers either online or in traditional retail outlets. Regardless of the sales form, access to information about products and their functional features is predominantly conducted digitally, through the manufacturer's website, blogs, social media, or video sharing platforms. Increasingly, this approach applies not only to promotional materials but also to user manuals and other product-related documents that customers need to receive. Though these activities might seem insignificant on their own, when multiplied by millions of products sold, they collectively substantially reduce a negative impact on the environment.

After the purchase, the customer starts using the acquired product, entering the phase of product usage monitoring, which closes the cycle.

7.3   Impact of digitalisation on sustainable products

Digitalisation, which has come to be known as the Industry 4.0, has disrupted traditional production and has changed the way in which people perceive goods, products, and services. Integrating digital technologies into systems encompasses environmental, social, economic, and technological aspects. This range of issues includes both benefits and difficulties in the process of finding a more sustainable world.

- Environmental impact

    - Reduction in carbon footprint: Digitalisation can lead to streamlined processes and efficient resource utilisation, lowering the overall carbon footprint of production.
    - Reduction of energy consumption: Controlling energy consumption by digital technologies such as Green IoT leads to avoiding wasteful energy in production processes (Paliwoda et al., 2023).
    - Minimising waste: Digital technologies enable precise monitoring and control over production processes, leading to minimisation of waste generation and more effective recycling and reuse of materials.
    - Preservation of natural resources: By optimising production processes through digitalisation, industries can reduce the extraction of raw materials, helping to preserve natural resources and biodiversity.

- Social impact

    - Job creation and workforce development: Digitalisation may automate certain tasks but also creates opportunities for new types of jobs, particularly in fields such as data analysis and software development.
    - Enhanced worker safety and well-being: Digital technologies can improve workplace safety by automating hazardous tasks and providing real-time monitoring of environmental conditions. Additionally, remote work opportunities may enhance work-life balance for employees.
    - Wider access to education: Technologies such as eLearning can increase learning outcomes and the quality of education, supporting students who "want relevant, mobile, self-paced, and personalised content in their learning process", it is also much cheaper and faster than traditional learning (Anggraini & Handayani, 2021).
    - Access to information about products: Digitalisation can facilitate transparency in supply chains, enabling consumers to make informed choices about sustainably produced goods and promoting fair trade practices.

- Economic impact

    - Increased efficiency and competitiveness: Digitalisation optimises production processes, reduces operational costs, and enhances product quality, improving the competitiveness of businesses in the global market.
    - Growth of sustainable industries: The adoption of digital technologies leads to the growth of new industries focused on renewable energy, eco-friendly materials, and circular economy models.

- Market expansion and consumer demand: As consumers become more environmentally conscious, there is a growing demand for sustainable products. Digitalisation allows businesses to reach a broader market and adapt to evolving consumer preferences (Elding & Morris, 2018).

- Technological impact

  - Integration of IIoT and data analytics: IoT devices embedded in production equipment can collect data on energy usage, resource consumption, and product performance, enabling better decision-making for sustainable practices.
  - Advancements in renewable energy integration: Digitalisation facilitates the integration of renewable energy sources such as solar and wind power into production processes, reducing reliance on fossil fuels and lowering greenhouse gas emissions.
  - Development of smart, eco-friendly materials: Digital technologies like CAD and simulation software enable the design and testing of innovative materials with reduced environmental impact, such as biodegradable plastics or sustainable composites.

Digitalisation is rapidly transforming the landscape of sustainable production, enabling industries to adopt eco-friendly practices and improve efficiency. However, this transition comes with its own set of challenges that need to be addressed for successful integration. Technical and technological challenges are a large part of them. Implementing sustainable production practices alongside the IoT may be very complex and require careful planning and execution. Financial constraints may be the next challenge. Budget issues hinder the adoption of sustainable practices in production processes. Additionally, insufficient funding for training personnel exacerbates the problem and the scarcity of skilled workers with expertise in digital technologies further complicates matters. The next issue is scalability. Scaling up sustainable production practices while managing the complexity of digital systems presents a formidable challenge. Ensuring the scalability of eco-friendly technologies to meet increasing production demands requires innovative solutions (Paliwoda et al., 2023).

Other challenges are related to information security and data protection. Ensuring data security and privacy in production processes is paramount. Protecting sensitive data from cyber threats requires introducing special security procedures. Another difficulty is ensuring the interoperability of the digital environment. It is crucial to establish standardised protocols for communication between production devices to ensure seamless integration and interoperability between devices from various manufacturers is vital for optimal performance.

Additionally, a significant challenge associated with digitalisation is the increased energy consumption related to data storage. As industries become more data-driven, the energy required to store, process, and manage vast amounts of data can have a negative environmental impact, potentially offsetting some of the gains made through sustainable practices. Addressing this issue requires the development of more energy-efficient data centres and storage solutions.

7.4   Conclusions and future perspectives

In today's fast-changing world, it's important to find better ways to enhance efficiency and simplify work processes and production methodologies. Digitisation is a key contributor to this need. With environmental degradation and rapid technological developments, we must pay particular attention to developing digital sustainability strategies throughout the product life cycle. Achieving sustainability in products and production processes requires a multidisciplinary approach, integrating digital technologies, innovative design, and a commitment to responsible resource management.

Digitalisation is a key enabler of sustainable production and offers possibilities for creating products that are not only innovative but also environmentally friendly and socially responsible. By leveraging technologies such as the IoT, AI, and BDA, companies can optimise product design, manufacturing processes, and lifecycle management to minimise environmental impact and maximise efficiency. The integration of digital technologies into production practices presents numerous benefits. From reducing carbon footprint and energy consumption to fostering job creation and market expansion, digitalisation has the potential to revolutionise the way we produce and consume goods. However, the journey towards sustainable production through digitalisation is not without its challenges. Technical complexities, financial constraints, scalability issues, and concerns regarding data security and interoperability pose significant risks that must be addressed. Overcoming these challenges requires collaboration, innovation, and a commitment to continuous improvement.

Looking ahead, the future of sustainable production lies in continued innovation and collaboration across industries, academia, and governments. To realise the full potential of digitalisation in sustainable products and sustainable production, investments in research and development, infrastructure, and workforce training are necessary. Governments can incentivise sustainable practices through policy frameworks, tax incentives, and regulatory measures. Education and awareness also play a crucial role in promoting sustainable consumption and production patterns.

References

Anggraini, R., & Handayani, Y. 2021. *Journal of Digital Education, Communication, and Arts* (DECA). 4, 2. https://doi.org/10.30871/deca.v5i01.2942

Cifuentes-Faura, J. 2022. European Union policies and their role in combating climate change over the years. *Air Quality, Atmosphere and Health*, 15(8), 1333–1340. https://link.springer.com/article/10.1007/s11869-022-01156-5

de Angelis, R., Howard, M., & Miemczyk, J. 2018. Supply chain management and the circular economy: Towards the circular supply chain. *Production Planning and Control*, 29(6), 425–437. https://doi.org/10.1080/09537287.2018.1449244

Elding, C., & Morris, R. W. 2018. Digitalisation and its impact on the economy: Insights from a survey of large companies. Economic Bulletin Boxes, 7. https://econpapers.repec.org/article/ecbecbbox/2018_3a0007_3a4.htm

Iacovidou, E., Velis, C. A., Purnell, P., Zwirner, O., Brown, A., Hahladakis, J., Millward-Hopkins, J., & Williams, P. T. 2017. A pathway to circular economy: Developing a conceptual framework for complex value assessment of resources recovered from waste. *Journal of Cleaner Production*, 2017, 166, 910–938.

IEC 62430: 2019. *Environmentally Conscious Design (ECD) — Principles, Requirements and Guidance*. International Organization for Standardization, Geneva.

Ingemarsdotter, E., & Dumont, M. 2022. August 12. Why the circular economy and LCA make each other stronger. *Circular economy Life cycle assessment*. Retrieved from: https://pre-sustainability.com/articles/the-circular-economy-and-lca-make-each-other-stronger, accessed on 31.03.2024.

ISO /TS14072#: 2014. *Environmental Management—Life Cycle Assessment—Requirements and Guidelines for Organizational Life Cycle Assessment*. International Organization for Standardization, Geneva.

ISO14006 #:2020. *Environmental Management Systems – Guidelines for Incorporating Eco-Design*. International Organization for Standardization, Geneva.

ISO14040 #:2009. *Environmental Management—Life Cycle Assessment—Principles and Framework*. International Organization for Standardization, Geneva.

Llompart, M., Sanchez-Prado, L., Pablo Lamas, J., Garcia-Jares, C., Roca, E., & Dagnac, T. 2013 Jan. Hazardous organic chemicals in rubber recycled tire playgrounds and pavers. *Chemosphere*, 90(2), 423–431. https://doi.org/10.1016/j.chemosphere.2012.07.053. Epub 2012 Aug 22. PMID: 22921644

Lopez-Galvez, N., Claude, J., Wong, P., Bradman, A., Hyland, C., Castorina, R., Canales, R. A., Billheimer, D., Torabzadeh, E., Leckie, J. O., & Beamer, P. I. 2022 Feb. Quantification and analysis of micro-level activities data from children aged 1–12 years old for use in the assessments of exposure to recycled tire on turf and playgrounds. *International Journal of Environmental Research and Public Health*, 19(4), 2483. https://doi.org/10.3390/ijerph19042483. PMID: 35206675; PMCID: PMC8879270

Lorang, S., Yang, Z., Zhang, H., Lü, F., & He, P. 2022. Achievements and policy trends of extended producer responsibility for plastic packaging waste in Europe. *Waste Disposal and Sustainable Energy*, 4(2), 91–103. https://link.springer.com/article/10.1007/s42768-022-00098-z

Mao, J., Li, C., Pei, Y., & Xu, L. 2018. Implementation of a circular economy. *Circular Economy and Sustainable Development Enterprises*. https://doi.org/10.1007/978-981-10-8524-6_9

Muralikrishna, I. V., & Manickam, V. 2017. Life cycle assessment. In *Environmental Management: Science and Engineering for Industry*, 57–75. https://doi.org/10.1016/B978-0-12-811989-1.00005-1

Paliwoda, B., Górna J., Biegańska M., & Wójcicki K. 2023. Application of Industrial Internet of Things (IIoT) in the packaging industry in Poland. *Logforum*, 19(1), 5. https://doi.org/10.17270/J.LOG.2023.787

PBL Netherlands Environmental Assessment Agency. 2017. Circular economy: Measuring innovation in the product chain (Policy Report). José Potting, Marko Hekkert, Ernst Worrell, & Aldert Hanemaaijer. The Hague. Retrieved from https://www.pbl.nl/

sites/default/files/downloads/pbl-2016-circular-economy-measuring-innovation-in-product-chains-2544.pdf, accessed on 31.03.2024.

Sphera. 2024. Sustainability and Regenerative Design. Retrieved from: https://www.spherasostenible.com/our-blog/regenerative-design, accessed on 28.03.2024.

UNEP, Life Cycle Initiative. 2015. Guidance on organizational life cycle assessment. Technische Universität Berlin (TU Berlin) and Kogakuin University

Winz, R., Yu, L. L., Sung, L.-P., Tong, Y. J., & Chen, D. (2023). Assessing children's potential exposures to harmful metals in tire crumb rubber by accelerated photodegradation weathering. *Scientific Reports*, 13(13877). https://doi.org/10.1038/s41598-023-38574-z

Zhu, Z., Liu, W., Ye, S., & Batista, L. 2022. Packaging design for the circular economy: A systematic review. *Sustainable Production and Consumption*, 32, 817–832. https://doi.org/10.1016/j.spc.2022.06.005

8

WASTE MANAGEMENT IN SUSTAINABLE DEVELOPMENT

Magdalena Muradin

8.1 Introduction

Waste management (WM) appears to be one of the most important concerns these days, which is a consequence of population growth, consumerism, and a linear approach to industrialisation. The volumes of generated waste tend to increase in every part of the globe, causing environmental, financial, and health problems. WM often creates a challenge according to the WM hierarchy, for raw material management, water and soil pollution, or cross-border transportation. WM is one of the biggest problems in modern society especially in terms of resources scarcity. How long will we be capable of satisfying the needs of current and future generations using a linear economic model?

In 2008, the European Commission published the Waste Framework Directive, which includes the current waste hierarchy. This hierarchy prioritises waste prevention over other WM options, such as preparing for reuse; recycling; other recovery methods, such as energy recovery; and disposal (Directive 2008/98/EC). In 2016, the waste hierarchy was included as 12th of the 17 Sustainable Development Goals (SDGs) defined as 'sustainable consumption and production'. In the context of sustainable development, waste prevention should be a key path of action. One way to implement this model is to adopt the principles of a circular economy (CE). This approach involves keeping raw materials and products in circulation for as long as possible and therefore reducing waste generation. One of the pioneers of the CE concept is the sailor Ellen McArthur, and the foundation she launched is the cornerstone for the whole CE concept (Ellen MacArthur Foundation, 2015a, 2015b). Just in practice, CE focuses on business models which ensure waste minimisation. However, its basis is to reduce the excessive use of primary raw materials while maintaining economic growth. Thus, CE is first and foremost

DOI: 10.4324/9781032710693-9

an economic model adapted to the changing political-social-economic-climatic conditions around the world, enabling sustainable development primarily in the area of raw material management. One of the operating models solely involves the conversion of waste generated during the production and use of goods into zero-impact resources for the production of secondary raw materials, which are then transformed into marketable goods. However, it should be borne in mind that an idealistic CE model would assume a completely closed material and energy cycle in the economy, which from the thermodynamic principles point of view has no right to exist (Muradin, 2022). Therefore, we can only achieve circularity to a certain extent, which leads to a situation of continuous closing the loop rather than a complete and full closure. The socio-economic development of society cannot take place in isolation from development in the field of WM (Liviu et al., 2021).

CE is still primarily seen from the perspective of increasing recycling levels and the reuse of materials and products. The aim of this chapter focus on discussing different business models of WM within the CE and identifying the most feasible sustainable solutions in accordance with the waste hierarchy, and with particular emphasis on processes such as recycling or waste-to-energy. The monitoring and measurement of sustainable WM is also worth of emphasising. This chapter focuses especially on the packaging and organic WM. The concept of industrial symbiosis (IS) in WM and the resulting benefits to create sustainable products will also be an important issue addressed in this chapter.

8.2   Circular economy concept

The first fundamental element of European Union (EU) policy and legislation for sustainable WM was the implementation of the waste hierarchy set out in the EU Waste Framework Directive (Directive 2008/98/EC). The purpose of the developed hierarchy in the form of an inverted pyramid was to do more than just regulate waste and protect the environment and human health; it aimed to increase resource and raw material efficiency as well. It was the first step towards the development of a CE. In 2018, the CE concept was anchored in the Directive amending Directive 2008/98/EC on waste (Directive (EU) 2018/851). WM was transformed into materials management in order to pursue sustainability, increase energy efficiency, provide new economic opportunities, and increase the Union's competitiveness compared to other countries in the world. CE has become a valid business model for keeping resources in circulation for as long as possible, reducing waste generation, and making materials management more sustainable. In fact, current WM should be seen as the management of economically valuable raw materials that provide material for the production of further goods and a potential source of revenue. The whole concept of CE has evolved over the years with successive publications, and a huge number of definitions have emerged to define the concept (Kirchherr et al., 2023). The International organisation for standardisation is developing an international standard on CE because there is still no framework within which we can move when implementing CE in an organisation (Muradin & Foltynowicz, 2019). It is argued that the basic principle of CE is to use the R-framework (Kirchherr et al., 2023; Singh et al., 2020). In the literature, there are widely described both basic 3 R-reduce-reuse-recycle and much more elaborate 9 R or even 38 R-imperatives models (Kirchherr et al., 2023; Reike et al., 2018). CE business models are primarily focused on implementing solutions rather than creating new products (Kirchherr et al., 2018; Lieder, Asif & Rashid, 2017). Some authors argue that CE is an appropriate tool for sustainable development (Sauvé, Bernard & Sloan, 2016; Ghisellini & Ulgiati, 2020). Others, on the other hand, argue that only in certain areas the implementation of CE principles can contribute to positive changes in the economy towards sustainability, but the close relationship between CE and SD is not clear (Geissdoerfer et al., 2017). This can be explained by the fact that CE can only be linked to a few of the 17 SDGs given by the United Nations in its report 'Transforming our World: The 2030 Agenda for Sustainable Development' (United Nations [UN], 2015). The strongest link can be found between CE and the Sustainable Consumption and Production Goal (SDG12). Geissdoerfer et al. argued that CE is a conditional concept for sustainability requiring primarily changes in the creation of value chains (Geissdoerfer et al., 2017). Therefore, one of the proposed development models is the creation of an IS allowing, for example, the conversion of waste into valuable raw materials necessary for further production. However, the limitations of CE do not affect its suitability for changing WM in a sustainable manner. Achieving environmental sustainability is highly dependent on effective WM, while effective WM is based on a CE and the concept of recycled waste as a potential future resource (Ranjbari et al., 2021).

The advantage of CE is that it can be used in two different technical and biological cycles (Ellen MacArthur Foundation, 2015a, 2015b). This means that CE principles can be applied to the management of both biological materials directly derived from biomass and so-called technical materials, that is, synthetic or mineral materials with the potential to remain in a closed system for production, recovery, and reuse (Panchal, Singh & Diwan, 2021).

8.3   Sustainable waste management in circular economy

Due to the morphological diversity of waste, its proper management raises serious difficulties and risks in both technological and economic spheres. Basically, WM includes all activities from collection and transport to treatment processes leading to the final disposal of waste (Aghbashlo, Tabatabaei & Hosseinpour, 2018; Rajaeifar et al., 2017). We call waste everything that the holder wants to get rid of. Therefore, we can classify waste according to residential, commercial, industrial, or institutional activities (Halkos & Aslanidis, 2023). In the whole life cycle of a product, waste is the last element of the product, the so-called end-of-life (EoL); however, proper WM is essential for the development of a sustainable and waste-free environment (Aghbashlo et al., 2019). Since waste is a plentiful source of secondary raw materials used for production processes, CE places this phase of the product life cycle very high in the new system development strategy. The linear produce–use–dispose model of the economy gained momentum, and the main goal of producers became to sell as many products as possible. Used products in the form of waste landed primarily in landfills. In addition, some manufacturing companies introduced patterns related to the 'planned obsolescence' of products, which consisted of the manufacturer planning the end of a product's life faster than expected due to its normal use through some hidden modifications (Antikainen & Valkokari, 2016). The changes that have taken place over the years in policy and legislation, especially in the EU, within the framework of WM have led to actions to reduce their impact on the environment, that is, sustainable waste management (SWM), which is based on waste hierarchy and enables all activities but in such a manner to prevent them from remaining in the nature and polluting the environment as well as affecting human health (Nelles, Gruenes & Morscheck, 2016). CE, on the other hand, takes us to a higher level of WM, it is no longer waste but reusable materials. And although Bilitewski argues that a key reason for the development of CE is to prevent waste and pollution, CE should be a much higher model than simply serving WM (Bilitewski, 2012). Above all, a model to reduce the consumption of primary raw materials, to extend the circulation of raw materials in the economy and to contribute to economic growth and enterprise development without a linear model. Thus, WM can only be a part of the whole CE model to meet the above demands. WM is intended not only to generate savings and revenue but also to steer businesses towards more circular and zero waste practices (Ranjbari et al., 2021).

8.4   Waste management business models

In this section, different types of business solutions supporting SWM will be presented. Current WM options based on the waste hierarchy will be discussed, as well as systems implemented under EU legal standards and business models that take into account the CE. It can be assumed that SWM will be based on the hierarchy established in Directive 2008/98/EC and that all EU countries base their waste prioritisation policies according to this order. On the other hand, if we consider that currently WM is only an element of raw material management in CE then the most appropriate approach would be the 6R methodology (Bradley et al., 2018). The links between SWM and CE are shown in Figure 8.1. In the case of CE, all 6R activities relate to the circulation of secondary materials, while in management according to the waste hierarchy, energy recovery processes are acceptable, although less preferred.

Selected WM models are discussed below: avoidance, reuse, and recycling, which build on the CE package of the EU, should be encouraged (European Commission, 2015). Processes such as energy recovery, or landfilling, although included in the WH concept of SWM, do not contribute to keeping raw materials in the economy for as long as possible. Energy recovery, although a better option than landfilling in the light of the CE, should be progressively eliminated

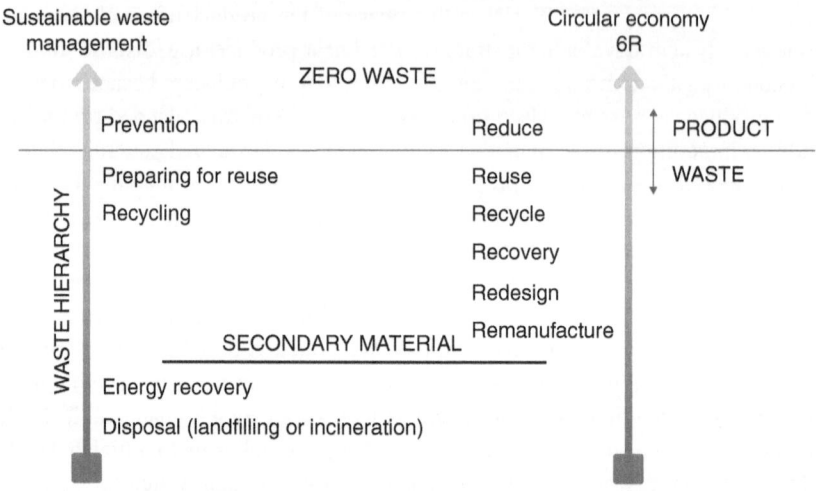

FIGURE 8.1 Relationship between sustainable waste management resulting from the waste hierarchy and circular economy in the 6R model.

Source: Own elaboration.

as a WM process. The incineration of waste to recover energy takes place at high temperature. Waste can be a very effective source of energy as in Denmark, where waste incineration covers about 5% of electricity demand and 20% of heat demand (Astrup, 2011). In some cases, incineration is the only recovery route due to the impossibility of separating the different fractions of waste with very different compositions. Therefore, new intelligent technologies are needed to enable the best possible separation of the different fractions and to catch recyclable materials. AI and progressive digitalisation can come to the rescue.

The waste hierarchy seems to focus mainly on recovery and recycling strategies, while the CE and overall policy should give highest priority to prevention (de Sadeleer, Brattebø & Callewaert, 2020).

8.4.1 Prevention and reduction

Waste prevention should be at the heart of SWM. The awareness of all stakeholders along the supply chain is important here: producers as well as consumers. Producers' actions can take place in relation to final products released onto the market. In that context, prevention already takes place at the source by extending the life cycle of products, producing good quality or repairable products, producing spare parts, or carrying out regular servicing and repairs (Table 8.1).

Producers can also act to streamline and increase the efficiency of production by eliminating waste generated in manufacturing processes, for example, through the implementation of new circular business models, resource efficiency measures, or appropriate by-product management. A by-product is defined as materials or

TABLE 8.1 Waste prevention practices

At the product manufacturer	Production in returnable packaging
	Good quality products
	Repairable products
	Production of interchangeable parts
	Thoughtful purchasing
	Provision of software updates for older versions of electronic devices
	Implementation of circular business models
	Efficient waste management
	Circularisation of production residues
	Management of by-products
	Sustainable ecodesign
At consumers	Buying products in returnable packaging
	Reducing consumption
	Buying reusable products
	Renting rarely used equipment
	Using equipment with a longer useful life (e.g., rechargeable batteries rather than batteries).
	Repairing rather than buying new

Source: Own elaboration.

substances that result from a production process, but the production of which was not the main purpose of that process. The concept of 'by-product' was introduced in Article 5 of Directive 2008/98/EC of the European Parliament and of the Council of 19 November 2008 on waste.

The prevention, or avoidance, of waste is a fundamental principle of the CE. However, it looks different from the consumers' and from the producers' perspective (Table 8.1).

Consumers are mainly required to reduce their consumption, change their habits, including food waste, and be considerate in their purchasing decisions. For producers, action should already focus on the product design stage. Design should be a tool to make significant, systemic changes, and shape our future (Bürdek, 2015). Circular design helps to fundamentally change the approach to creating new products and services. It shows how to maintain the value of raw materials, materials, and products over time by managing their life cycle loops. It is necessary to start creating a completely different type of product, to move from a 'fast' approach to a 'slow' approach, for example, as is the case in the 'fast fashion' industry (Rocha, Antunes & Partidário, 2023):

- will be long lasting due to their durability or by creating a symbolic attachment to the user;
- can be reconfigured to fulfil different functions;
- will be designed to facilitate repair and renovation.

After a design process, changes in the operational phase concerning the implementation of circular business models or efficient WM should be discussed.

Waste prevention requires the interaction of all stakeholders in a value chain. The action of public institutions also makes an important contribution to the development of CE in WM (Saidani et al., 2017).

8.4.1.1   End of waste criteria

One of the legal actions a producer can take in the management of post-consumer waste is to achieve end-of-waste status for a certain type of waste. End-of-waste criteria are set out in Article 6(1) and (2) of the Waste Framework Directive (2008/98/EC) defining when waste ceases to be waste and becomes a product or a secondary raw material. This refers to waste that has undergone a recovery operation (including recycling) and which (Directive 2008/98/EC):

- can be widely used for other purposes, for example, cullet;
- a market or demand exists for the substance or object in question;
- the use is legal;
- the use will not lead to overall environmental adverse or human health impacts.

End-of-waste criteria are developed by the Joint Research Centre (JRC) considering a high level of environmental protection and environmental and economic benefits. The end-of-waste criteria do apply to raw materials after the recycling process, that is, no waste is initially avoided, this only happens after they have been processed and prepared for reuse; however, some of these materials do not lose their original properties in the process and can be reused several times, such as glass cullet, which can help companies to develop circular business models, remove unnecessary administrative burdens, and standardise the entire WM system.

Criteria have already been developed for:

- iron, steel, and aluminium scrap (Council (EU) No. 333/2011);
- cullet (EU) No. 1179/2012);
- copper scrap (EU) No 715/2013);

while criteria proposals for plastics and textiles are under preparation.

8.4.2   Reuse

Reuse encompasses a range of non-destructive activities leading to the further use of solid materials without altering their physico-chemical state, leading to an extended lifetime of these materials (Cooper & Gutowski, 2017). Again, we can distinguish here at least two situations at the level of the consumer, when we have an already used finished product, and at the level of the producer when we are talking about, for example, the reuse of post-production materials. In the first case, reuse options refer to the exchange of products, resale of used products (e.g., on Internet exchanges), and reuse after remanufacturing. Many of these activities were known long before the CE concept was introduced, for example, the sale of used cars. Nowadays, it has become easier thanks to the development of the internet and various platforms for exchanging used products including the sale of second-hand clothing.

Reuse will have a completely different dimension in companies. A suitable example is the described case of a furniture company (Hartini et al., 2021), which uses waste boards and pieces of wood to create new tables or remanufactures defective products by restoring as many parts as possible to their original condition or form without losing functionality.

8.4.3   Recycling

The global demand for raw materials is continuously increasing due to rising standards of living, greater demand for material resources, and the development of industrialisation. Demand is increasing for biomass resources, fossil fuels, metals, and non-metallic minerals even with sustainable development measures and regulations being implemented. The UN's Global Resources Outlook 2024 report indicates that the COVID-19 pandemic had temporarily slowed the growth of resource extraction, but growth rates have since recovered (Bruyninckx et al., 2024). It is estimated that resource extraction will reach 106.6 billion tonnes in 2024, up from 30 billion tonnes in 1970, and is expected to be as high as 195 billion tonnes by 2060 (OECD, 2019). Still, about 60% of mined minerals and metal ores are currently used as primary anthropogenic resources (Krausmann et al., 2017). This is why waste recycling processes are so important for the economy, as they can meet future material demands without relying on virgin raw materials. Europe is leading the way in implementing policy incentives for waste sorting and recycling. One of the more important principles being implemented in the EU is extended producer responsibility (EPR), which will be discussed in Section 8.4.3.1. Increasing recycling levels faces many difficulties such as:

- Ensuring that appropriate technologies allowing material recovery without incurring losses each time materials are reintegrated into a new production and consumption cycle.
- Ensuring significant energy inputs due to the necessity of converting waste into secondary raw materials in processes that often change their physical and chemical properties and dealing with emissions to the environment, which is not ideal from a CE point of view.
- The limited availability of waste materials for recycling, resulting not only from decentralised waste collection but also from the long life of certain products. On one hand, such an approach may run counter to the reduce and reuse principle

in the CE; on the other hand, it results in a reduction in the availability of these materials and thus the need for increased extraction of virgin raw materials. However, decentralisation of recycling (the collection and dismantling of components and the extraction of raw materials for recycling do not take place on such a large scale as in the case of the extraction of virgin raw materials) not only requires extensive infrastructure for the collection and transport of materials to recycling facilities but also depends on consumer preferences (Krook & Baas, 2013).

- Technical difficulties when dismantling complex products such as electronic equipment, photovoltaic panels, electric cars, and many others. These appliances are characterised by the small size of the individual modules and the complex assembly of many small components, which, if disassembly is not standardised, can lead to significant losses of valuable components and materials. It is estimated that approximately 60% of gold and more than 80% of palladium are lost during the pre-processing of mobile phones in preparation for the recycling process (Bacher, Mrotzek & Wahlström, 2015).

In recycling, special attention should be paid to its role in the CE. Recycling is not an ideal solution but should only be a step in the transformation to achieve a closed loop in both the technical area and the biological cycle. Two theories can be recalled here: the so-called rebound effect or Jevon's Paradox and the complexity theory. The rebound effect assumes that although technological progress can reduce human impact on the environment, for example, by increasing resource and energy efficiency, in the contrary such technologies can increase economic growth through demand. This drives towards reduce and reuse as an activity that promotes sustainable consumption and production versus recycling, which can contribute according to this rebound effect to stimulate consumption and use of much more resources in the future (Bacher, Mrotzek & Wahlström, 2015).

Complexity theory originated as an attempt to describe complex physical non-equilibrium systems but is increasingly being used to describe processes in living nature. It indicates that complex systems cannot be designed from the top down nor can their behaviour be predicted. A complex system designs itself through the interaction or evolution of its components. In contrast, and most importantly for CE and recycling, the system expansion entails an exponential increase in the resources needed to sustain its operation. Thus, the more complex the system, the more innovative the technologies and processes, the more resources will be consumed until they are completely exhausted, making sustainability ultimately unattainable.

8.4.3.1   Extended producer responsibility

EPR, considered as one of the main WM policy instruments to support the implementation of the European waste hierarchy, was introduced by the OECD in 1994 as an element of environmental policy according to which producers are responsible for the final life-cycle stage of their products, not consumers. This means that producer responsibility for a product does not terminate at the production stage but is extended to the other stages of the life cycle and, in particular, to the final disposal stage. The EPR is a model approach setting out how to deal with a product that is a product of human activity (Muradin, 2023). In EU law, the principle of EPR was introduced by Article 8 of the Waste Framework Directive (Directive 2008/98/ EC). The EPR implies that the producer is not only financially responsible but also organisationally responsible at the stage of the final management of the product when it becomes waste and is complementary to the overarching 'polluter pays' principle underpinning EU environmental policy. It also remains directly linked to the principle of sustainable development and SWM.

The EPR pursues a double objective:

- to make producers responsible for the costs of WM instead of the public (i.e., taxpayers).
- to support eco-design of products to minimise these costs and the associated environmental damage.

The EPR covers packaging waste, electrical and electronic equipment, batteries, accumulators, and EoL vehicles. However, with regard to EPR, a ban on the marketing of single-use products such as cutlery, plates, stirrers, straws, cosmetic sticks, and polystyrene beverage containers and cups was further introduced in 2021 (Directive (EU) 2019/904).

In an EPR system, each actor in the product lifecycle value chain should have clearly defined responsibilities (Table 8.2).

TABLE 8.2 Responsibility of value chain stakeholders in an EPR scheme

Stakeholder	Responsability
Producer	• to cover the costs necessary to achieve waste management objectives and targets, including waste prevention and reduction; • the provision of a service and spare parts catalogue; • the use of secondary raw material in production; • financing and organising a waste collection and recycling system; • to inform consumers through appropriate product labelling; • to support information campaigns on responsible consumer behaviour;
Consumer	• appropriate collection and separation of waste generated; • financing the management of mixed waste;
Legislator	• to establish an effective regulatory regime in the form of compulsory levies, and also incentives; • to establish an effective financial and non-financial instruments.

Source: Based on Directive (EU) 2019/904.

When implementing the EPR, special attention should be paid to other activities that go directly beyond the WM framework but relate to the implementation of EU solutions in line with broad environmental policy, sustainable development, and the direction of the global economy (Muradin, 2023). These activities should primarily address the eco-design stage, that is, designing products in such a way that they have the least possible environmental impact throughout their life cycle by introducing design solutions for product repair and reuse, recyclability and recovery of the raw materials from which they are made.

It is recognised that EPR can significantly help to contribute to existing EU waste targets as it provides the basis for efficient EoL collection, or more effective re-use and recycling of collected products. Maitre-Ekern (2021) claims that EPR policies do not provide waste prevention and do not support the development of CE. It is expected that EPR will link products and waste by providing incentives for producers to improve product design, ultimately save WM costs (Maitre-Ekern, 2021). It can be argued that the EPR counteracts the CE's basic idea of waste avoidance with regard to reduce and reuse if only by negligibly promoting corrective action in waste and product legislation (Saari et al., 2019). EPR allows for increased recycling rates and the implementation of modern recovery technologies, thereby mitigating the effects of WM; however, it unfortunately does not reduce the amount of waste generated.

The impact of EPR on the product design stage is still negligible (Maitre-Ekern, 2021). It is more cost-effective to incur recycling costs than to change the process line and product design changes. In order to encourage change and implementation of CE rules, it is not only possible to focus on increasing recycling rates but to move further towards harmonising regulations to include minimum requirements for product and material design and the origin of these materials. It could also introduce differentiate or modulate the EPR fees aiming to guide producers towards more responsibility for their products (Watkins et al., 2017).

8.5   Monitoring and measurement

Progress on CE as well as SWM requires adequate monitoring and the identification of indicators to enable these measurements. There is a significant number of indicators for measuring circularity at global, regional, and micro levels for different companies. These indicators can address different areas of sustainability. Some indicators in line with WH and CE are presented in Table 8.3.

In the case of SWM, indicators should be chosen that illustrate the increase in efficiency of activities in this area. Measurements can be made using indicators from different perspectives, for example, to diagnose and detect problems (Guerrero & Erbiti, 2004) or to assess the efficiency of WM (Bringhenti, Zandonade & Günther, 2011; Hotta et al., 2014).

A good measure would be to link CE indicators directly to indicators on the waste hierarchy as an overarching prerogative to enable SWM. Pires and Martinho

TABLE 8.3 Waste management indicators in line with WH and CE

Indicators in line with the waste hierarchy		Indicators in line with the 6Rs model for the circular economy	
Amount of waste generated	Waste generated (kg)/time (year)	Percentage of recycling rate of all waste	Waste recycled (kg)/ generation of waste (kg) × 100
Percentage of waste reused in manufacturing processes	Waste reused in manufacturing processes (kg)/ waste generated (kg) × 100	Percentage of recycling rate of plastic waste	Plastic waste recycled (kg)/ generation of waste (kg) × 100
Amount of waste recycled	Waste recycled (kg)/ time (year)	Percentage of recycling rate of paper and paperboard	Paper and paperboard waste recycled (kg)/ generation of waste (kg) × 100
Amount of waste treated in other recovery process	Waste treated in other recovery process (kg)/time (year)	Generation of waste per EUR (kg/ EUR)	Generation of waste/ revenues
Amount of energy generated in a recovery process	Energy (GJ)/time (year)	Percentage of circular material use (CMU) rate	Secondary materials (kg)/material consumption (kg) × 100
Amount of waste disposed of or landfilled	Waste disposed or landfilled (kg)/ time (year)	Percentage of generation of waste per material consumption	Generation of waste (kg)/materials consumption (kg) × 100
Amount of hazardous waste generated	Hazardous waste (kg)/time (year)	By-products delivered or sold	By-products (kg)/ time (year)
By-products generated	By-products generated (kg)/ time (year)		

Source: Own elaboration.

(2019) point to the lack of indicators directly related to WH implementation. Research is mainly focused on building indicators to measure performance in terms of waste collected and costs incurred, thus seeking methods and tools that can improve and optimise treatment activities (Rodrigues, Martinho & Pires, 2016). In the case of WM, the focus is on ensuring the measurement of the effectiveness of different technologies, or WM models, and on the collection of data and information, particularly in relation to recycling (Price & Joseph, 2000). The recycling rate itself is a quantifiable indicator. It often indicates a movement up the hierarchy away from disposal but does not illustrate the actual level of waste generated, only the amount that has been recycled (Price & Joseph, 2000). More difficult is to measure indicators related to waste minimisation or avoidance.

The development of standardised, uniform indicators as a kind of benchmark is still lacking in the case of CE, especially for company-level indicators. Many of these indicators are still in the pilot phase (Walker et al., 2018). Furthermore, it is up to the organisations themselves to select the indicators they wish to use to measure their circularity from a large pool of indicators available and proposed by various authors, such as:

- material circularity indicator on the amount and intensity of circular and regenerative flows at the micro level (Ellen MacArthur Foundation, 2015a, 2015b).
- an indicator to quantify the CE performance of different plastic waste treatment options based on the technical quality of the plastic waste stream (Huysman et al., 2017).
- a value-based resource efficiency indicator for assessing resource efficiency based on resources returned at EoL and then re-marketed (Di Maio et al., 2017).
- a circularity index based on the recirculation of economic value to total product value (Linder & Williander, 2017).

8.6   Sustainable packaging waste management

Sustainable packaging WM systems are crucial to achieve the sustainable development and ambitious goals sets by European legislation. The reduction of waste by establishing a closed-loop recycling of food packaging materials is a key element of the EU's CE action plan (Franz & Welle, 2022). There are many challenges with sustainable management packaging waste including the whole management chain beginning with waste collection and separation processes, preparation for recycling, recycling technologies, etc. It stems from the heterogeneity of the materials and their chemical–physical properties, as well as from the heterogeneity of policies, lack of knowledge, autonomous recycling targets, etc.

It has to be highlighted that reduce and reuse of waste in CE concerns rather the management of materials than waste. Reduce and reuse should be considered as priority. The development of production technologies and product design should focus on minimalisation of packaging or using reusable packaging rather than on recycling waste already generated. Recycling should be intended only as a transitional phase. However, there are two different approaches to recycling of packaging waste that can be considered in the light of CE:

- closed loop recycling
- open loop recycling

The idea of a closed loop is to return the product for use in creating more products of the same kind, for example. A glass bottle factory is responsible for glass cullet

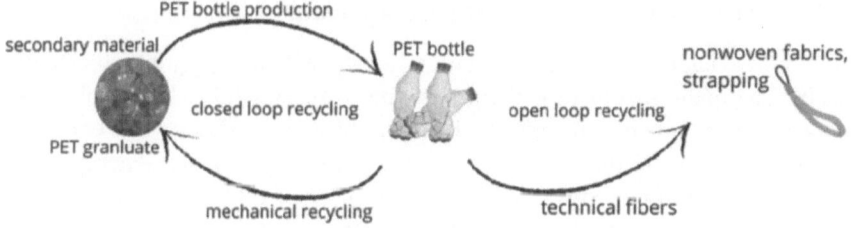

FIGURE 8.2 The example of closed and open loop recycling of PET bottle.

Source: Own elaboration.

collection and to return the recycled material to the same production process of bottles. On the other hand, the open loop allows for the use of recycled materials in the industry to manufacture different products of another kind (Nakatani & Moriguchi, 2014), such as using glass cullet as a construction material. The closed loop should support to maintain the quality of the material at the same level without major weight losses and ensuring that this material circulates in the economy for as long as possible.

The modelled example in polyethylene terephthalate (PET) bottle recycling is the closed loop supply chain created for value recovery in business-to-business sector (Figure 8.2).

The cooperation of stakeholders in a supply chain can improve to preserve the material in technosphere and avoid the extraction of raw materials. The production of bottles starts from amorphous, virgin PET (Lee et al., 2024). Then this life cycle of PET bottle is a business as usual until the EoL stage. Bottles are collected at drop-off points in shops and delivered to one recycling centre to be cleaned from contaminants. High-quality recycled PET is transformed into flakes (granules) and then pellets before entering the bottle production processes, which is called closed-loop recycling. On the other hand, the open-loop recycling provides other applications usually downcycling processes. The polymer production turned out to be the most environmentally damaging what can substantially reduce by replacing 50% of the total amount of virgin PET used with rPET (Ingrao & Wojnarowska, 2023). In view of the sustainable development, transport should also be considered. That is why the collecting and recycling points in closed loop should be located at the same place as the bottle production factories to diminish the impact of transport. This leads to cooperation between recyclers and producers, which can be based on IS rules, which is one of the business models acknowledged in CE. Municipal packaging WM is primarily characterised by open-loop recycling, especially with single-use products that do not return to the supply chain of the factories that produced them. On the other hand, mixing similar waste with different properties impedes recycling.

8.7 Sustainable organic waste management

As with packaging WM, dealing with organic waste also requires to establish an IS. The main goal of IS is the promotion of sustainability by establishing a cooperation between companies with similar business profiles while enhancing competitiveness. Surplus resources in many forms (by-products, energy, water, etc.) generated in one industrial process can be redirected as a new input into another process by one or more other companies and create competitive benefits for all participants (Zago et al., 2021). IS can also be compared to biological processes, in which organisms associate in a "mutually beneficial relationship" (Neves et al., 2019). In order to make the IS more environmentally efficient, cooperation should take place between entities located close to each other. This reduces, for example, the environmental impact of transport and can lead at the same time to increased economic benefits. In CE, the anaerobic digestion and composting processes establish the restoration cycle to biomass (Sherwood, 2020). Returning biomass to the cycle as ferment or compost improves biomass production at least by returning the biogenic elements and minerals to the ground (Sherwood, 2020). A relevant example is a model of an industrial and agricultural cluster that can consist of a biogas plant, dairy cows and pig farms, maise and rape crops cultivation, a spirits distillery, an oil mill, and a biofertiliser factory, all run by several producers associated in one cooperative. This IS model improves the use of resources and enables the biomass recirculation what results in waste generation reduction, by-products management, and minimises emissions into the environment (Muradin, 2022). All operations are carried out directly between cluster participants in a closed loop. The IS improves the competitiveness of the cluster with regard to other individual entities on the market (Muradin, 2022).

Adequate examples can also be found in the literature. One of them presents the possible architecture of IS in Sodankylä, Finland (Haq, Mazumder & Kalamdhad, 2020). The IS includes combined heat and power woodchip fuel-based plants, a biogas reactor, greenhouse tomatoes farm, fish farm, and several insect farms. According to SWM approach, the cooperation provides a significant waste reduction (Haq, Mazumder & Kalamdhad, 2020).

References

Aghbashlo, M., Tabatabaei, M., & Hosseinpour, S. (2018). On the exergoeconomic and exergoenvironmental evaluation and optimization of biodiesel synthesis from waste cooking oil (WCO) using a low power, high frequency ultrasonic reactor. *Energy Conversion and Management*, *164*, 385–398.

Aghbashlo, M., Tabatabaei, M., Soltanian, S., & Ghanavati, H. (2019). Biopower and biofertilizer production from organic municipal solid waste: An exergoenvironmental analysis. *Renewable Energy*, *143*, 64–76.

Antikainen, M., & Valkokari, K. (2016). A framework for sustainable circular business model innovation. *Technology Innovation Management Review*, *6*(7), 5–12.

Astrup, T. (2011). Optimal utilization of waste-to-energy in an LCA perspective. *Waste Management, 31*, 572–582.

Bacher, J., Mrotzek, A., & Wahlström, M. (2015). Mechanical pre-treatment of mobile phones and its effect on the Printed Circuit Assemblies (PCAs). *Waste Management, 45*, 235–245.

Bilitewski, B. (2012). The circular economy and its risks. *Waste Management, 32*(1), 1–2.

Bradley, R , Jawahir, I. S., Badurdeen, F., & Rouch, K. (2018). A total life cycle cost model (TLCCM) for the circular economy and its application to post-recovery resource allocation. *Resources, Conservation and Recycling, 135*, 141–149

Bringhenti, J. R., Zandonade, E., & Günther, W. M. R. (2011). Selection and validation of indicators for programs selective collection evaluation with social inclusion. *Resources, Conservation and Recycling, 55*(11), 876–884.

Bruyninckx, H., Hatfield-Dodds, S., Hellweg, S., Schandl, H., Vidal, B., Razian, H., Nohl, R., Marcos-Martinez, R., West, J., Lu, Y., Miatto, A., Lutter, S., Giljum, S., Lenzen, M., Li, M., Cabernard, L., Fischer-Kowalski, M., Kulionis, V., Oberschelp, C., & Pfister, S. (2024). *Global Resources Outlook 2024: Bend the trend-pathways to a liveable planet as resource use spikes*. United Nations Environment Programme. https://wedocs.unep.org/20.500.11822/44901 (accessed on 02.04.2024).

Bürdek, B. E. (2015). *History, Theory and Practice of Product Design*. Birkhauser Varlag, p. 14.

Cooper, D. R., & Gutowski, T. G. (2017). The environmental impacts of reuse: A review. *Journal of Industrial Ecology, 21*(1), 38–56.

de Sadeleer, I., Brattebø, H., & Callewaert, P. (2020). Waste prevention, energy recovery or recycling—Directions for household food waste management in light of circular economy policy. *Resources, Conservation and Recycling, 160*, 104908.

Di Maio, F., Rem, P. C., Baldé, K., & Polder, M. (2017). Measuring resource efficiency and circular economy: A market value approach. *Resources, Conservation and Recycling, 122*, 163–171.

Directive 2008/98/EC of the European Parliament and of the Council of 19 November 2008 on waste and repealing certain Directives (Text with EEA relevance) (Dz. Urz. UE 2008 L 312/3).

Directive (EU) 2018/851 of the European Parliament and of the Council of 30 May 2018 amending Directive 2008/98/EC on waste (Text with EEA relevance).

Directive (EU) 2019/904 of the European Parliament and of the Council of 5 June 2019 on the reduction of the impact of certain plastic products on the environment (Text with EEA relevance).Ellen MacArthur Foundation (2015a). Material Circularity Indicator (MCI). https://www.ellenmacarthurfoundation.org/material-circularity-indicator (accessed on 02.04.2024).

Ellen MacArthur Foundation (2015b). Circular economy report – Delivering the circular economy. https://www.ellenmacarthurfoundation.org (accessed on 02.04.2024).

European Commission (2015). Communication from the commission to the European Parliament, the Council, the European Economic and Social Committee and the Committee of the Regions. Closing the loop – An EU action plan for the circular economy. Brussels, 2.12.2015, COM(2015) 614 final.

Franz, R., & Welle, F. (2022). Recycling of post-consumer packaging materials into new food packaging applications—Critical review of the European approach and future perspectives. *Sustainability, 14*, 824.

Geissdoerfer, M., Savaget, P., Bocken, N. M. P., & Hultink, E. J. (2017). The circular economy – A new sustainability paradigm? *Journal of Cleaner Production*. https://doi.org/10.1016/j.jclepro.2016.12.048

Ghisellini, P., & Ulgiati, S. (2020). Managing the transition to the circular economy. In *Handbook of the Circular Economy*. M. Brandão, D. Lazarevic, & G. Finnveden (eds.) (pp. 491–504). Edward Elgar Publishing.

Guerrero, E., & Erbiti, C. (2004). Indicadores de sustentabilidad para la gestión de los residuos sólidos domiciliarios. Municipio de Tandil, Argentina. *Revista de Geografía Norte Grande, 32*, 71–86.

Halkos, G. E., & Aslanidis, P. S. C. (2023). New circular economy perspectives on measuring sustainable waste management productivity. *Economic Analysis and Policy, 77*, 764–779.

Haq, I., Mazumder, P., & Kalamdhad, A. S. (2020). Recent advances in removal of lignin from paper industry wastewater and its industrial applications – A review. *Bioresource Technology, 312*, 123636.

Hartini, S., Wicaksono, P. A., M D Rizal, A., & Hamdi, M. (2021). Integration lean manufacturing and 6R to reduce wood waste in furniture company toward circular economy. *IOP Conference Series: Materials Science and Engineering, 1072*(1), 12067. https://doi.org/10.1088/1757-899X/1072/1/012067.

Hotta, S., Yamao, T., Bisri, S. Z., Takenobu, T., & Iwasa, Y. (2014). Organic single-crystal light-emitting field-effect transistors. *Journal of Materials Chemistry C, 2*(6), 965–980.

Huysman, S., De Schaepmeester, J., Ragaert, K., Dewulf, J., & De Meester, S. (2017). Performance indicators for a circular economy: A case study on post-industrial plastic waste. *Resources, Conservation and Recycling, 120*, 46–54.

Ingrao, C., & Wojnarowska, M. (2023). Findings from a streamlined life cycle assessment of PET-bottles for beverage-packaging applications, in the context of circular economy. *Science of the Total Environment, 892*, 164805.

Kirchherr, J., Piscicelli, L., Bour, R., Kostense-Smit, E., Muller, J., Huibrechtse-Truijens, A., & Hekkert, M. (2018). Barriers to the circular economy: Evidence from the European Union (EU). *Ecological Economics, 150*, 264–272.

Kirchherr, J., Yang, N. H. N., Schulze-Spüntrup, F., Heerink, M. J., & Hartley, K. (2023). Conceptualizing the circular economy (revisited): An analysis of 221 definitions. *Resources, Conservation and Recycling, 194*, 107001.

Krausmann, F., Wiedenhofer, D., Lauk, C., Haas, W., Tanikawa, H., Fishman, T., … & Haberl, H. (2017). Global socioeconomic material stocks rise 23-fold over the 20th century and require half of annual resource use. *Proceedings of the National Academy of Sciences, 114*(8), 1880–1885.

Krook, J., & Baas, L. (2013). Getting serious about mining the technosphere: A review of recent landfill mining and urban mining research. *Journal of Cleaner Production, 55*, 1–9.

Lee, C., Jang, Y.-C., Choi, K., Kim, B., Song, H., & Kwon, Y. (2024), Recycling, material flow, and recycled content demands of polyethylene terephthalate (PET) bottles towards a circular economy in Korea. *Environments, 11*(2). https://doi.org/10.3390/environments11020025

Lieder, M., Asif, F. M., & Rashid, A. (2017). Towards circular economy implementation: An agent-based simulation approach for business model changes. *Autonomous Agents and Multi-Agent Systems, 31*, 1377–1402.

Linder, M., & Williander, M. (2017). Circular business model innovation: Inherent uncertainties. *Business Strategy and the Environment, 26*(2), 182–196.

Liviu, M., Razvan, P., & Marcuta, A. (2021). The relationship between the circular economy and sustainable waste management in European Union. *Journal of Business Administration Research, 4*(1), 37–44.

Maitre-Ekern, E. (2021). Re-thinking producer responsibility for a sustainable circular economy from extended producer responsibility to pre-market producer responsibility. *Journal of Cleaner Production, 286,* 125454.

Muradin, M. (2022). Circular models for sustainable supply chain management. In *Sustainable Products in the Circular Economy. Impact on Business and Society*. M. Wojnarowska, M. Ćwiklicki, & C. Ingrao (eds.) (pp. 1–18). New York, Routledge, 294 s., ISBN 9781003179788. DOI:10.4324/9781003179788

Muradin, M. (2023). Legal aspects of environmental protection in Poland. In *Environmental Management*. A. Matuszak-Flejszman (ed.), Poznań, Poznań University of Economics Publishing House (in polish), 13–34.

Muradin, M., & Foltynowicz, Z. (2019). The circular economy in the standardized management system. *Amfiteatru Economic, 21*(13), 871–883. https://doi.org/10.24818/EA/2019/S13/871

Nakatani, J., & Moriguchi, Y. (2014). Time-series product and substance flow analyses of end-of-life electrical and electronic equipment in China. *Waste Management, 34*(2), 489–497

Nelles, M., Gruenes, J., & Morscheck, G. (2016). Waste management in Germany–development to a sustainable circular economy? *Procedia Environmental Sciences, 35,* 6–14.

Neves, A., Godina, R., Azevedo, S. G., Pimentel, C., & Matias, J. C. O. (2019). The potential of industrial symbiosis: Case analysis and main drivers and barriers to its implementation. *Sustainability, 11*(24), 7095. https://doi.org/10.3390/su11247095

OECD. Global Material Resources Outlook to 2060: Economic Drivers and Environmental Consequences 2019. https://doi.org/10.1787/9789264307452-en

Panchal, R., Singh, A., & Diwan, H. (2021). Does circular economy performance lead to sustainable development?–A systematic literature review. *Journal of Environmental Management, 293,* 112811.

Pires, A., & Martinho, G. (2019). Waste hierarchy index for circular economy in waste management. *Waste Management, 95,* 298–305.

Price, J. L., & Joseph, J. B. (2000). Demand management–a basis for waste policy: A critical review of the applicability of the waste hierarchy in terms of achieving sustainable waste management. *Sustainable Development, 8*(2), 96–105.

Rajaeifar, M. A., Ghanavati, H., Dashti, B. B., Heijungs, R., Aghbashlo, M., & Tabatabaei, M. (2017). Electricity generation and GHG emission reduction potentials through different municipal solid waste management technologies: A comparative review. *Renewable and Sustainable Energy Reviews, 79,* 414–439.

Ranjbari, M., Saidani, M., Esfandabadi, Z. S., Peng, W., Lam, S. S., Aghbashlo, M., ... & Tabatabaei, M. (2021). Two decades of research on waste management in the circular economy: Insights from bibliometric, text mining, and content analyses. *Journal of Cleaner Production, 314,* 128009.

Reike, D., Vermeulen, W. J., & Witjes, S. (2018). The circular economy: New or refurbished as CE 3.0?– Exploring controversies in the conceptualization of the circular economy through a focus on history and resource value retention options. *Resources, Conservation and Recycling, 135,* 246–264.

Rocha, C. S., Antunes, P., & Partidário, P. (2023). Design for circular economy in a strong sustainability paradigm. *Sustainability, 15*(24), 16866.

Rodrigues, S., Martinho, G., & Pires, A. (2016). Waste collection systems. Part B: Bench-marking indicators. Benchmarking of the Great Lisbon Area, Portugal. *Journal of Cleaner Production, 139,* 230–241. https://doi.org/10.1016/j.jclepro.2016.07.146

Saari, U., Fedoruk, M., Iital, A., Moora, H., Klöga, M., & Voronova, V. (2019). An overview of the problems posed by plastic products and the role of extended producer responsibility in Europe. *Journal of Cleaner Production, 214*, 550–558.

Saidani, M., Yannou, B., Leroy, Y., & Cluzel, F. (2017). How to assess product performance in the circular economy? Proposed requirements for the design of a circularity measurement framework. *Recycling, 2*(1), 6.

Sauvé, S., Bernard, S., & Sloan, P. (2016). Environmental sciences, sustainable development and circular economy: Alternative concepts for trans-disciplinary research. *Environmental Development, 17*, 48–56.

Sherwood, J. (2020). The significance of biomass in a circular economy. *Bioresource Technology, 300*, 122755.

Singh, R. K., Kumar, A., Garza-Reyes, J. A., & de Sá, M. M. (2020). Managing operations for circular economy in the mining sector: An analysis of barriers intensity. *Resources Policy, 69*, 101752.

United Nations [UN] (2015). Transforming our world: Agenda for for sustainable development 2030. Resolution adopted by the General Assembly on 25 September 2015 (A/RES/70/1).

Walker, S., Coleman, N., Hodgson, P., Collins, N., & Brimacombe, L. (2018). Evaluating the environmental dimension of material efficiency strategies relating to the circular economy. *Sustainability, 10*(3), 666.

Watkins, E., Gionfra, S., Schweitzer, J. P., Pantzar, M., Janssens, C., & ten Brink, P. (2017). *EPR in the EU Plastics Strategy and the Circular Economy: A Focus on Plastic Packaging*. Institute for European Environmental Policy (IEEP).

Zago, P. C., Cotrim, S. L., Leal, G. C. L., Galdamez, E. V. C., & Ferreira, M. A. (2021). An industrial symbiosis method applied to waste management. *Environmental Engineering and Management Journal, 20*(6), 905–915.

9

THE ROLE OF CORPORATE SOCIAL RESPONSIBILITY FOR SUSTAINABILITY OF PRODUCTS

Agata Matarazzo, Massimo Riccardo Costanzo and Pierluigi Catalfo

9.1   Introduction

Strategic management scholars have focused attention on whether the fit between environment, ethic and firm strategy motivates acquisition behaviour; in fact environmental and social uncertainty affects whether firms select to acquire or opt for other cooperative means. Strategic management, often called 'policy' or nowadays simply 'strategy,' is about the direction of organizations, and most often, business firms. Strategy is the creation of a unique and valuable position, involving a different set of activities. Strategic position emerges from distinct sources: serving few needs of many customers, serving broad needs of few customers or serving broad needs of many customers in a narrow market (Porter, 1996). Strategic management as a field of inquiry is firmly grounded in practice and exists because of the importance of its subject. The strategic direction of business organizations is at the heart of wealth creation in modern industrial society (Rumelt, Schendel & Teece, 1991). Social responsibility, therefore, understood as a new strategic approach in management system that induces companies to take a path aimed at enhancing the relationship with the stakeholder network and at creating social, environmental and sustainable development innovation, leads companies to seize growing entrepreneurial and market opportunities. Its key themes are linked to working conditions and standards that determine innovative ways of strategic management: from the environmental one, to social and gender equity, to the availability and management of data, to new technologies and anti-corruption policies.

The theme of corporate social responsibility (CSR) is now consolidated mainly in economic and managerial contexts (Arezzo, D'Amico & Randone, 2017; Asongu, 2007) It consists of all the socially responsible behaviour of an undertaking vis-à-vis its shareholders, and, more generally, the community,

DOI: 10.4324/9781032710693-10

generally adopted voluntarily and regardless of compliance with legal obligations (Thiengnoi & Afzal, 2009). Companies' growing awareness of this concept has led to the creation of a multiplicity of application tools (Romolini, 2007)]. In the 1960s, the notion of CSR was definitively affirmed and, thanks to numerous academic contributions, a rapid growth in its relevance to the business activity was witnessed, thus involving the birth and development of two different forms of thought (Del Baldo, 2008). One of these establishes the existence of a link between the responsibilities of managers and their social power, due to the fact that the opinions and actions of businessmen influence society. Avoiding such obligations leads to a progressive corrosion of power and negative consequences on business activity (Ezekiel et al., 2013). The other form of thinking, on the other hand, attempts a more articulated definition of CRS, emphasizing the importance of the expectations of the community in which the company operates and the consequent social role in increasing the well-being of these communities. Finally, the fundamental stage of the historical evolution of the concept dates back to the 1980s (Bisio, 2015). It was characterized by the conception and emergence of the first empirical studies on social performance generated by the implementation of CSR actions (Hagedoorn & Schakenraad, 1994). The social function of the enterprise, therefore, goes beyond the strictly economic aspects.

It follows that the possibility of success of the company is closely related to the inclusion of socio-economic aspects, which are hardly linked to the income result of the organization. Such elements may be the distinctive source of the competitive aspect of an undertaking (Brown, de Jong & Lessidrenska, 2009). The explosion of the concept of CSR in recent years is due to a crisis of values that has marked the industrial system at a global level (Bagnoli, 2010). In fact, there has been an increasing number of bad business behaviour, all based on the search for easy profits at the expense of the quality of life of citizens, their safety, their health and social cohesion. The concept of CSR has been formalized for the first time in the Green Paper presented in 2001 under the title Promoting an European Framework for CSR. On this occasion, the Committee provided the unambiguous definition CSR by describing it as the voluntary integration of the social and ecological concerns of companies in their business operations and in their relations with the parties interested (Gulati, 1998). The implementation of a CSR strategy is a crucial component for the competitiveness of an enterprise and it is something that must be coordinated by the enterprise itself. This involves the implementation by the company of policies and methodologies that integrate social, environmental, ethical, business operations and its main strategy, accomplishing everything in close collaboration with stakeholders. The aim of this chapter is to know and deepen the theme of CSR, analysing its typical features and international tools useful to companies to undertake behaviours considered socially responsible. The analysis of this component, fundamental for companies, not only identifies the theoretical aspects of social responsibility but also aims to understand how these aspects can be put into practice within the different organizations.

9.2   Corporate social responsibility tools

Over the years, the growing formalization of CSR has been undertook through the design and development of organizational tools, which allowed companies to communicate and to bring to the attention of the commitments made, in the field of social responsibility, all its stakeholders. The proliferation of the following instruments has been encouraged by the evolution and dissemination of transparency issues and the principles of social responsibility, the attention of the political class to the impact of companies in society and the affirmation of an increasingly rooted sensitivity to the issues of sustainable development (Casotti, 2005). The first stage of the formalization process of the CSR was marked by the setting of reference values, the mission and objectives of the company (Bruni & Sarti, 2009). Subsequently, they moved to the design and development of tools able to communicate and monitor the objectives set in order to see that the commitments made to its stakeholders have been respected (Burgers, Hill & Kim, 1993). The tools that are processed during the research are:

- ethical code: represents a social contract between the company and its stakeholders by publicly announcing that it is aware of its citizenship obligations, which has developed business policies and practices consistent with these obligations and which is able to implement them through appropriate organizational structures (Cannella & Hambrick, 1993).
- social report: is a fundamental tool to improve social and industrial relations and strengthen in the community the positive perception of the actions of the socially responsible company, increase the legitimacy and consensus of the community (De Colle & Gonella, 2003).
- standard SA 8000: it analyses a wide range of disputes including child and forced labor, health and safety, the right to collective bargaining, discrimination, remuneration and management systems related to supply chain verification, external communication management and other business policies (Farné, 2012).
- standard ISO 26000: is an international standard that provides guidelines on CSR and Organizations (Castka & Balzarova, 2008; Hahn, 2013).
- accountability 1000: is an intentional membership standard prepared in 1999 by the International Council of the Institute of Social and Ethical Accountability (ISEA), formed by companies (www.csqa.it).

9.2.1   Code of ethics

By code of ethics, we refer to the official document that outlines the values and principles guiding the company's daily activities and the responsibilities it assumes towards the various stakeholders (employees, suppliers, customers, the public administration, shareholders, the financial market, etc.). The adoption of a code of ethics allows the company to promote the rules of conduct and moral obligations that each member is required to respect, seeking to prevent and prohibit any unlawful behaviour by those individuals who act in the name and on behalf of the organization (www.randstad.it/knowledge360/employer-branding/ecco-cose-il-codice-etico-aziendale).

This document represents a support to the Organization and Management Model outlined in Legislative Decree 231/2001, but its scope is broader and aims to ensure that all activities are developed with the utmost respect for the environment and socio-economic aspects. The structure of the code of ethics is defined according to the type of company; in any case, it is possible to identify five general levels:

1 at the first level are the company's own ethical principles, to which each member must refer;
2 at the second level, one can find the principles and ethical standards that enable the management of relations between the company and all stakeholders;
3 at the third level, there are the ethical standards of conduct that allows to determine the correct conduct to be maintained towards stakeholders with reference to specific areas (e.g., protection of the person, protection of health, the principle of moral legitimacy);
4 at the fourth level, sanctions may be established as a consequence of violation and non-compliance with the rules contained in the code;
5 finally, procedures may be adopted to monitor the implementation of the code, possibly aimed at revising it.

The proposal of the code of ethics presupposes an appropriate knowledge of the interests and expectations that stakeholders have of the organization. An element of fundamental importance when drawing up ethical standards and principles is the involvement and internal sharing with stakeholders; in fact, efficient management of the interdependencies between the company and stakeholders generates mutual benefits, including greater productivity, economic growth, and better performance. In conclusion, it is reasonable to state that the code of ethics represents a kind of 'moral contract' between the company and the stakeholders, which can be relied upon both in the execution of normal day-to-day management and in the management of activities that can have a long-term impact on the organization (Herciu, 2016). For this reason, a direct involvement during the implementation of the code increases the sense of belonging of each stakeholder towards the company.

9.3 Sustainability reporting

By sustainability reporting (accounting, balance sheet, report), the company provides information on its sustainability strategy and policy, how to implement them and the criteria for evaluating the relevant performance measures (Strampelli, 2022).

The origin of sustainability reporting, as it is emerging at the European level, can be found in the Paris Agreement ratified by the EU in 2016, following the adoption in 2015 of the United Nations 2030 Agenda for Sustainable Development, in which it was agreed to limit global warming to no more than 2°C.

On 8 March 2018, the European Commission published a first Sustainable Finance Action Plan, with the aim of channelling more finance towards environmentally sustainable economic activities, in particular those activities that can play a crucial role in achieving a zero-emission and climate-resilient economy by 2050.

With the 2019 Green Deal, the European Commission adopted a set of proposals to transform the EU's climate, energy, transport and taxation policies to reduce net greenhouse gas emissions and achieve a 'green transition' to climate neutrality by 2050.

To finance the green transition, public and private financial flows must be directed towards sustainable economic activities. Therefore, it becomes necessary to know in depth how sustainable businesses are, so that investors – including banks – can make choices in deciding: what (which businesses) to invest in, to whom (which companies) to grant credit and what conditions to apply.

9.3.1   The corporate sustainability reporting directive

The need to express the value created by the company, the increased awareness of the impacts of economic activities, and the need to keep up with the new consumer purchasing dynamics have pushed companies towards non-financial reporting, especially large globalized and listed companies, which are confronted with increasingly attentive and demanding stakeholders. This is a reporting model that goes beyond traditional reporting and allows for a retrospective and prospective reconstruction of the value creation process undertaken by the company, expressing the links between the financial and non-financial aspects and reporting on a wide range of aspects, such as governance, resource allocation, strategies, future prospects, risk management, the relationship with the external environment and the use of capital (Quagli, Corsi & Trucco, 2021). Fundamental is to communicate indicators that are useful taking into account the specific circumstances, relevant, consistent and most pertinent with the metrics one has decided to use in the internal and risk assessment processes, and to communicate them in a reliable way between one financial year and the next one, in order to provide consistent information on progress and orientations (Atripaldi et al., 2022).

The purpose of non-financial reporting is to consider the company not as the sum of numbers and rules, whose objective is exclusively profitability, but to consider it as the result of a set of values, responsibilities, integrations and sharing (Strampelli, 2022). The corporate sustainability reporting directive (CSRD) requires certain specifically named organizations to draw up a sustainability report (or sustainability budget) that contributes to an understanding of how sustainable the company is and provides the basis for dialogue with the various stakeholders.

Through sustainability reporting, the company provides:

- all information on how developments in the field of sustainability influence and affect the company; an example is the effects of climate change on the business model. This is also referred to as financial materiality;
- all information on the effects the company itself has on the surrounding environment (system, conditions); for example, the effect of emissions from production processes on the air quality of local residents. This is referred to as impact materiality.

In jargon, they speak of 'dual materiality' when referring to these two perspectives as a whole: the impact on the enterprise and the impact of the enterprise.

The sustainability report/balance sheet should contain all material (i.e., relevant) information. Information is considered material/relevant when its omission or misrepresentation influences the user's assessment and judgement.

So a topic that is material: either only from a financial point of view, or only from an impact point of view, or from both perspectives (on the company and of the company) certainly becomes part of the sustainability report.

The CSRD applies to:

- large enterprises and public interest entities;
- small and medium-sized listed companies;
- non-EU companies with a net turnover of EUR 150 million in the EU and at least one subsidiary or branch in the EU.

Non-EU companies are obliged to prepare sustainability reports in the following cases:

- if the non-EU enterprise has a net turnover of more than EUR 150 million within the EU for two successive financial years;
- the non-EU company has a subsidiary that qualifies as a listed SME and/or has a branch with a net turnover of more than EUR 40 million for the previous financial year.

Companies that do not fall within the scope of CSRD (e.g., micro-enterprises even if they are listed) may still be affected by CSRD, for example, if they are part of the supply and/or value chain of a company obliged to report on sustainability – but with different limitations.

The obligation to draw up and make public the sustainability report will come into force in stages:

From the financial year, 1 January 2024, a CSRD sustainability report will be mandatory for companies that currently already prepare a non-financial statement – contained in the management report accompanying the annual financial statements – pursuant to the NFRD.

These are Public Interest Entities (PIEs, banks, insurance companies, listed companies) with more than 500 employees. The first CSRD sustainability reports will be public from the beginning of 2025.

From the financial year, 1 January 2025, mandatory CSRD sustainability reporting for large companies with a European legal form. Companies that are part of a group can, as stated below, benefit from an exemption.

From the financial year, 1 January 2026, mandatory CSRD sustainability reporting for listed SMEs.

As of the financial year, 1 January 2028, mandatory CSRD sustainability reporting for non-EU companies (without EU legal form).

Large companies will be required to prepare and publish a sustainability report for financial years starting in 2025. If a company has an 'interrupted' financial year (which, e.g., runs from 1 September 2024 to 31 August 2025), it will start preparing sustainability reporting for the financial year beginning 1 September 2025. In fact, the financial year starting on 1 September 2025 is – in the example – the first financial year starting from the effective date of the CSRD reporting obligation (Baumüller & Grbenic, 2021).

National legislators may extend the scope of application, for example, by obliging cooperative societies, foundations, associations, mutual insurance companies, municipalities, government institutions, pension funds, investment institutions, housing associations, and health and educational institutions – in the long run – to prepare sustainability reporting.

The integrated report should enable the achievement of a number of prevailing objectives:

- broaden and balance the set of performance indicators, usually focused on the short-term financial dimension, by incorporating also indicators with a medium-to long-term perspective;
- properly represent the interactions between financial and non-financial news, especially with regard to decisions that impact on the organization's long-term performance;
- meeting the information needs of investors;
- provide a comprehensive framework so that environmental and social factors are systematically internalized within management processes and the reporting system (Indelicato, 2014).

The three pillars of integrated reporting, regardless of the reference model adopted, can be summarized as follows:

- communicate the social, economic and environmental impacts of the company's activities to the various stakeholders, providing an overview of performance.
- to broaden and improve the possibilities of analysis, evaluation and choice of stakeholders, thus allowing a comparison of the results obtained over the years by the same company, or by different companies.

- pushing the company towards continuous performance improvement (Rimini, 2018; Cristiano & Ferraro, 2022).

In March 2021, as part of the revision of the non-financial reporting directive, the European Lab Project Task Force published the final report responding to the European Commission's request to the EFRAG (European Financial Reporting Advisory Group) for technical advice on the introduction of non-financial reporting standards in the European Union.

The document recommends the definition of European standards concerning, inter alia:

- the qualitative characteristics of information;
- the time horizons of reporting;
- the illustration of the principle of dual materiality, according to which companies are required to provide information on the impacts generated in the area of sustainability and the consequent effects on the corporate organization;
- the integration of financial and sustainability statements.

The standards should be structured on three levels:

1 general, applicable to all obliged parties;
2 sectoral, with specific indications for the economic sector to which it belongs;
3 individual, left to the freedom of individual companies to facilitate the disclosure of specific information.

These regulatory measures and international positions are the expression of a broad and articulated movement, which looks towards a new era of capitalism, much more compatible with the socio-natural context in which companies operate today. It is evident that the notion of 'non-financial information' is to be understood in a broad sense, and can also cover in an evolving way other potentially very interesting information such as corporate reputation. In other words, these legislative measures are especially important for the new avenues they are able to open up towards more advanced reporting (www.efrag.org).

9.3.2   Structure of the sustainability report

The contents of the sustainability report cannot be identified a priori, since they depend on the materiality analysis of the company, just as they cannot be generalized, since there are standards that must be adapted to the specific reality being reported (Mavellia et al., 2010; Morsing & Roepstorff, 2014). However, it is possible to identify basic contents present in the generality of sustainability reports. They are:

- the letter to stakeholders: is drafted by a member of top management (President, Vice-President, etc.).

- the methodological note: this describes the report's drafting principles, the stakeholder engagement process, the reporting boundary, the materiality analysis performed and the methods.
- the organization's profile: this gives a general picture of the company, that is, its history, mission and vision, values, its activities, products and services offered, the stages of organization of production, its geographical presence and size, as well as any membership of associations.
- governance and strategy: this section includes information on the structure and role of the organization's governing bodies, and illustrates the sustainability strategy undertaken.
- reporting on materiality aspects: this can be articulated in different ways, for example, by type of impact (economic, social, environmental) or by stakeholder (customers, suppliers, employees, local community, etc.).
- economic performance: here we provide the most detailed representation of the value generated and distributed, making it possible to quantify how much wealth has been produced and how it is distributed to both internal and external stakeholders: shareholders and employees, suppliers, public administration, territory and community.
- environmental performance: this section deals specifically with the impacts the company generates on living and non-living natural systems, including water, soil, air and ecosystems. Information can be included on resources used (such as raw materials, energy, water) and outputs produced (e.g., discharges, emissions, waste).
- social performance: this part includes information and indicators on employment and working conditions, health and safety at work, training and education, diversity and equal opportunities (Minutiello & Tettamanzi, 2022).

9.3.3   The materiality matrix

In the context of sustainability reporting, 'materiality' is used as a threshold to consider all those topics that have a strong influence on stakeholder assessments and decisions and a high relevance in terms of economic, social and environmental impacts, and are therefore important enough to be reported on. This makes materiality a highly malleable concept, as there are different interpretations of it and no objective process for its determination, which is why there is an increasing demand for standardization. In this regard, both EFRAG and ISSB (International Sustainability Standards Board) are trying their hand at creating global reporting standards, which adopt a 'dual materiality' approach introduced by CSRD. This approach provides a structured representation of the environmental and social risks to which the company is exposed and the effects that the company's activities have on sustainability factors.

Dual materiality consists of defining:

- how the published data may influence the company's own market position, performance and development ('inside-out' perspective);

- the impact of the company's activities on the external context, society and the environment (outside-in perspective) (Baldi, 2023; Confindustria, 2020).

In the materiality analysis process, the identification and assessment of stakeholder categories is the first step to be taken, so one could start with a stakeholder mapping exercise, highlighting which stakeholders are internal and which are external to the company. For the purposes of assessing the issues relevant to stakeholders, the next step would be to involve them by administering a questionnaire to a number of individuals congruent to homogeneously represent the stakeholder categories identified.

9.4   Global Reporting Initiative indicators

Companies that draw up the declaration on non-financial information are required to follow a reference reporting standard and/or a completely autonomous methodology explicitly indicated in the DNF. This choice will affect the performance indicators to be adopted for the evaluation and monitoring of the company's activities. It is possible to identify several initiatives at a global level that set themselves the objective of defining guidelines for a reporting tool that also includes ethical and sustainability values. Among these, it is worth highlighting the Global Reporting Initiative (GRI), a non-profit association founded in 1997, which was born from the famous agreement between the US governmental organization called Coalition for Environmentally Responsible Economies (CERES) in collaboration with the United Nations Environment Programme (UNEP) (Yosifova & Petrova-Kirova, 2022).

The aim of this initiative was to develop, through a multi-stakeholder process, a standard capable of making non-financial reporting more effective by integrating the social, economic and environmental dimensions of sustainability itself, and to make it usable worldwide. The aim is to make the non-financial report as close as possible to the levels of rigour, comparability, credibility and verifiability proper to financial statements (Tettamanzi & Minutiello, 2021). This reporting model, proposed by GRI, has been widely adopted by companies, becoming the main reference worldwide for the realization of sustainability reporting, being based on a broad spectrum of parameters relating to the economic, social and environmental dimension that allows the communication of the company's global performance during the financial year to be brought back to a single document, in a triple bottom line perspective. The GRI standards enable every company – whether large or small, private or public – to understand and report its impacts in a comparable and (www.asvis.it) credible way, thereby increasing transparency on its contribution to sustainable development. In addition to organizations, the standards are of particular importance to many stakeholders (institutions, investors, etc.).

The standards were conceived as a modular set that could be used immediately and offer a complete picture of a company's main issues, their impacts and how these are managed.

In particular, based on recent revisions of the Standards, three different types of standards can be distinguished:

- the Universal Standards, which apply to all organizations and incorporate human rights reporting and environmental due diligence, in line with inter-governmental expectations. These standards offer the highest level of transparency for impacts on the economy, environment and people (www.nexteco.it/gri-standards-cosa-cambia-dal-2023/).
- sector Standards, which allow reporting on the most significant specific impacts specific to the sector of the company in question. They are designed to increase the quality, completeness and consistency of the information reported: they describe the contexts within which the different sectors develop, suggest their (likely) material themes and for each one list the specific aspects to be reported.
- topic Standards, adapted for use with the Universal Standards, list disclosures related to a particular topic.

The GRI guidelines are constantly evolving in order to improve their applicability for each type of organization. The first GRI guideline (G1) was released in 2000 with the aim of improving the comparability and quality of information at a global level on economic, environmental and social impacts, ensuring greater transparency on the part of organizations (Hedberg & von Malmborg, 2003). In 2002, the second version of the GRI guidelines was issued (G2), which was subsequently updated in 2006 (G3) and again in 2011 (G3.1) (Heimann, 2008). In 2013, G4 was introduced, which reinforced corporate disclosure of business risks. In 2016, GRI defined universal standards (GRI 101, 102, 103) and specific standards (200, 300, 400) both applicable to all kinds of companies, small and large, both public and private, as they are based on a common and universal language to guide communication about their sustainable impact (García et al., 2019).

Specific standards, on the other hand, included many different standards that organizations could use:

- GRI 201 to GRI 207 (Economic): which covered economic performance, market presence, indirect economic impacts, anti-corruption practices, etc;
- GRI 301 to GRI 308 (Environmental): related to materials used, water, energy, biodiversity, waste, emissions, assessment of suppliers' environmental impacts;
- GRI 401 to GRI 419 (Social): concerning education, labour, forced labour, rights of indigenous peoples, respect for human rights, respect for local communities, supply chain, consumer protection, etc. (Rossi, Orelli & Del Sordo, 2018).

9.5   Dissemination and benefits

In recent years, sustainability has come to play an increasingly central role in the business processes and development projects of European institutions.

The implementation of a non-financial reporting process brings several advantages to companies. First of all, it contributes to breaking down the internal barriers of the organizations that adopt it, as it brings about a progressive rapprochement of the different corporate functions involved in the various reporting processes (finance, marketing, human resources, purchasing, legal area, etc.). In fact, such an approach favours a high level of internal collaboration, marked by a greater awareness of the impact that the individual decisions of one organizational unit cause on all the others (Khanna, Gulati & Nohria, 1998).

In particular, the adoption of an integrated report favours the analysis of the links between financial and non-financial performance, paving the way for a better understanding of the cause-and-effect relationships that link all company activities. Furthermore, it removes barriers between the company and the stakeholders, and between the various stakeholders themselves, who until recently were the recipients of specific documents addressed to them, representing a connecting element that synergistically concentrates the interests, perspectives and expectations of the various actors. It promotes dialogue, involvement and commitment of the entire stakeholder network, thanks also to the role of the Internet that allows for a high level of interaction. In addition, the attention paid to the environmental dimension through reporting processes, represents a bridge towards the affirmation of an increasingly comprehensive and global approach compared to mere environmental policy, as the company sees itself taking on ever broader responsibilities, on multiple fronts related not only to environmental protection but also to safety at work, gender equality, equity, social cohesion, reducing the level of corruption, etc. In other words, it has anticipated a broader orientation towards sustainability on the part of companies, enabling them to raise their corporate culture to an ever-increasing level of responsibility towards society and to increase their ability to adapt to factors of change, or rather to anticipate and exploit them in order to maximize results.

In addition, the emergence and spread of non-financial reporting documents among companies has certainly represented a revolution, as they constitute wide-ranging documents, autonomous in their drafting but complementary to the contents of the annual financial statements.

Another considerable advantage is that the establishment of an adequate non-financial disclosure system considerably reduces the phenomenon of 'greenwashing,' thus preventing companies from distorting the direction of investment flows by declaring the pursuit of specific sustainability policies, without such statements being followed by concrete actions (www.impresaprogetto.it/editorials/2021-1/rusconi; Moggi, 2016).

9.6   Conclusions

In this particular historical phase, no company can afford to resist external changes. The process of 'natural selection' of organizations due to climate change, globalization, environmental challenges, poverty alleviation and social hardship needs to be taken into account (Williamson, 1999). This is why non-financial reporting is an extraordinary tool available to companies. At the European level, the regulatory environment is constantly changing and companies are increasingly encouraged to adopt responsible behaviour. Moreover, directives, regulations and recommendations concerning the reporting of certain information by companies are increasingly numerous and varied. In the light of the above, non-financial information therefore represents a fundamental step in the evolutionary path that Europe is taking to improve the ability of companies to generate value and to communicate it in an increasingly integrated way for the benefit of all stakeholders inside and outside the company. For success, it is essential to involve corporate functions with different knowledge and skills by facilitating dialogue, to create awareness of the usefulness of environmental and social information, and to increase the effectiveness of controls to improve the usability and security of such information. Finally, it is necessary and beneficial for organizations to escape from the 'ideological' view of such developments, which should instead be seen as an evolution of the economic-productive system towards processes more compatible with the modern context (Bossut et al., 2021). It can therefore be said that the European Commission has taken a big step forward in the field of sustainability by implementing the CSRD Directive, as this directive will be an essential tool for highlighting how sustainability reports are essential for communicating how environmental and social policies contribute to value creation. Through the CSRD, the European Commission will finally be able to close the gaps in regulations on sustainability, and improve the completeness and reliability of sustainability reports within the EU, which will be imposed on approximately 50,000 large companies that play a strategic and fundamental role within the socio-cultural, territorial and economic contexts within each Country of the European community.

At the dawn of 2024, in a world marked by geopolitical rivalries, climate change, resource scarcity, inflation, and social and environmental crises, there is a growing need for clear rules, universally applied standards, greater transparency, and a reduction in areas where rules or regulations leave room for greenwashing. Otherwise actually the risk of a perpetual state of crisis will be high, and CSRD bodes well for a Europe in which organizations will be projected to invest in healthy future growth, human development and green technologies é (Arfò & Matarazzo, 2023).

Companies that focus some of their energies on implementing a proper environmental accounting system will gain an awareness that a green economy does not stifle economic development; rather, it promotes reinvestment in natural capital. Companies that do not adapt to the changes in sustainability reporting will run the risk of exiting the market, as they will no longer be evaluated only on their financial performance, but also on their sustainability performance, thus remaining one step behind in a world moving towards a sustainable world.

References

Arezzo C., D'Amico R., Randone S., 2017, *La responsabilità sociale oltre l'impresa*, Franco Angeli, Milano.

Arfò S., Matarazzo A., 2023, Corporate sustainability reporting directive as strategic management tool to apply corporate social responsibility in sustainable companies, Conference Proceedings XL Convegno Nazionale AIDEA 2023 L'aziendalismo crea valore! Il ruolo dell'accademia nelle sfide della società, dell'economia e delle istituzioni, October 2023, pp. 1–13.

Asongu J., 2007, *Strategic Corporate Social Responsibility in Practice*, Greenview Publishing Company, Vancouver.

Atripaldi E., Gila N., Musco A., Sauerwald U., 2022, *La diversità di genere nelle dichiarazioni non finanziarie delle banche d'Italia*, Bank of Italy, Rome, pp. 5–10.

Bagnoli, L., 2010, *Responsabilità sociale e modelli di misurazione*, Franco Angeli, Milano.

Baldi P., 17 January 2023, *The New Sustainability Reporting Directive (CSRD) and Audit of Non-Financial Data – The Concept of Materiality: Definitions and Future Developments*, ODCEC Roma Commissione ESG, Sviluppo Sostenibile e Corporate Reporting, Rome, pp. 6–8.

Baumüller J., Grbenic S., 1 January 2021, *Moving from Non-Financial to Sustainability Reporting: Analysing the EU Commission's Proposal for a Corporate Sustainability Reporting Directive (CSRD)*, International Accounting Group, Vienna, p. 379.

Bisio L., 2015, *Comunicazione aziendale di sostenibilità socio-ambientale*, G. Giappichelli, Torino.

Bossut M., Jürgens I., Pioch T., Schiemann F., Spandel T., Tietmeyer R., July 2021, *What Information Is Relevant for Sustainability Reporting? The Concept of Materiality and the EU Corporate Sustainability Reporting Directive*, Sustainable Finance Research Platform Policy Brief Augsburg, pp. 2–3.

Brown H., de Jong M., Lessidrenska T., 2009, The rise of the Global Reporting Initiative: A case of institutional entrepreneurship, *Environmental Politcs*, 27, 182–200.

Bruni D., Sarti M., 2009, *Rendicontazione e partecipazione sociale: dal bilancio sociale al bilancio partecipativo*, Franco Angeli, Milano.

Burgers W. P., Hill C. W. L., Kim W. C., 1993. A theory of global strategic alliances: The case of the global auto industry, *Strategic Management Journal*, *14*(6), 419–432.

Cannella A. A., Hambrick D. C., 1993, Effects of executive departures on the performance of acquired firms, *Strategic Management Journal*, *14*, 137–152.

Casotti A., 2005. *La responsabilità sociale delle imprese* (Vol. 5). Wolters Kluwer Italia, Milano.

Castka P., Balzarova, M. A., 2008, ISO 26000 and supply chains – On the diffusion of the social responsibility standard, *International journal of Production Economics*, *111*(2), 274–286.

Confindustria, 2020, *Confindustria per la sostenibilità*, Centro studi Confindustria, Rome, pp. 14–15.

Cristiano E., Ferraro O., 2022, *Le BCC e l'informativa non finanziaria*, Franco Angeli, Milano, pp. 13–17.

De Colle S., Gonella C., 2003, Corporate social responsibility: The need for an integrated management framework, *International Journal of Business Performance Management*, 5, 199–212.

Del Baldo M., 2008, *Corporate Social Rresponsibility e corporate governance: un nesso vincente nelle PM*, Bepress, Urbino.

Ezekiel S., Asemah Ruth A., Okpanachi, Leo O., Edegoh N., 2013, Business advantages of corporate social responsibility practice: A critical review, *New Media Mass Communication*, *180*, 45–54.

Farné S., 2012, Qualità sostenibile. *Strategie e strumenti per creare valore, competere responsabilmente e ottenere successo duraturo. Le norme ISO 26000, SA 8000, ISO 9004, ISO 14000*, Franco Angeli, Milano, pp. 183–185.

García I., Hussain N., Martínez J., Ruiz E., 2019, Impact of disclosure and assurance quality of corporate sustainability reports on access to finance, Corporate Social Responsibility and Environmental Management, *26*(4), 832–848.

Gulati R., 1998, Alliances and networks, *Strategic Management Journal*, *19*, 293–317.

Hagedoorn, J., Schakenraad, J., 1994, The effect of strategic technology alliances on company performance, *Strategic Management Journal*, *15*(4), 291–309.

Hahn R., 2013, ISO 26000 and the standardization of strategic management processes for sustainability and corporate social responsibility, *Business Strategy and the Environment*, *22*, 442–455.

Hedberg C., von Malmborg F., 2003, *The Global Reporting Initiative and Corporate Sustainability Reporting in Swedish Companies*, Wiley Interscience, Sweden.

Heimann, G., 2008, *Corporate Social Responsibility. Global Standards & Policies in Practice*, The Liberian International Ship & Corporate Registry. U.S.A.

Herciu M., 2016, ISO 26000 - An integrative approach of corporate social responsibility, *Studies in Business and Economics*, *11*, 73–79.

Indelicato F., 2014, *Report e Reporting integrato: verso un nuovo modello di bilancio*, *Equilibri*, *18*(1), 9–37

Khanna T., Gulati R., Nohria N., 1998, The dynamics of learning alliances: Competition, cooperation and scope, *Strategic Management Journal*, *19*(3), 193–210.

Mavellia N., Randazzo F., Tavernar E., Baroni E., Caprioli L., Chiacchio G., Dall'Anese R., Navassa A., Spoldi P., 2010, *La rendicontazione della sostenibilità*, ODCEC di Milano, Milan.

Minutiello V., Tettamanzi P., 2022, *ESG: Sustainability Report and Integrated Reporting*, Ipsoa, Milan, pp. 15–17.

Moggi S., 2016, *Il Sustainability Reporting nelle università*, Maggioli Editore, Santarcangelo di Romagna, pp. 31–36.

Morsing M., Roepstorff A., 2014, *CSR as Corporate Political Activity: Observations on IKEA's CSR Identity-Image Dynamics*, Springer, Denmark.

Porter M., 1996, What is strategy? *Harvard Business Review*, *74*(6), 1–20.

Quagli R., Corsi K., Trucco S., 2021, *Bilancio ed informativa economico-sociale*, Giappichelli Editore, Turin, pp. 77–82.

Rimini E., 2018, *I valori della solidarietà sociale nelle dichiarazioni non finanziarie*, Analisi Giuridica dell'Economia, *17*(1), 187–200.

Romolini A., 2007, *Accountability e bilancio sociale negli enti locali*, Franco Angeli, Milano.

Rossi F., Orelli R., Del Sordo C., 2018, *Integrated Reporting and Corporate Value*, Franco Angeli, Milan, pp. 207–210.

Rumelt R. P., Schendel D., Teece D., 1991, Strategic management and economic, *Strategic Management Journal*, *12*, 5–29.

Strampelli G., 2022, L'informazione non finanziaria tra sostenibilità e profitto, *Analisi Giuridica dell'Economia*, *21*(1), 148–149.

Tettamanzi P., Minutiello V., 2021, *Il bilancio di sostenibilità come strumento di rendicontazione aziendale*, Guerini Next, Milano, pp. 25–27.

Thiengnoi P., Afzal S., 2009, *A Comparative Analysis of CSR Strategies, Implementation and Outcomes*, Diva, Karlstand.

Williamson O. E., 1999, Strategy research: Governance and competence perspectives, *Strategic Management Journal, 20*, 1087–1108.

www.asvis.it

www.csqa.it

www.efrag.org

www.impresaprogetto.it/editorials/2021-1/rusconi

www.nexteco.it/gri-standards-cosa-cambia-dal-2023/

www.randstad.it/knowledge360/employer-branding/ecco-cose-il-codice-etico-aziendale.

Yosifova D., Petrova-Kirova M., 5 November 2022, *The New EU Corporate Sustainability Reporting Framework in the Context of GRI Standards*, Economic and Social Development: Book of Proceedings, 85–95.

10

MARKETING SUSTAINABLE PRODUCTS

Magdalena Wojnarowska, Erica Varese and Szymon Jarosz

10.1   Introduction

Contemporary entrepreneurs are increasingly aware of the need to prioritize environmental protection in response to shifts in consumer preferences within the market for goods and services. As environmental consciousness grows, consumers are showing a greater inclination to choose products that are healthy, safe, and produced in accordance with environmentally friendly practices. This burgeoning interest of society in adopting a pro-ecological lifestyle is compelling companies to employ ecological marketing strategies on a more extensive basis. Consequently, individuals are finding it increasingly important to embrace environmentally friendly behaviours.

As consumer awareness of environmental issues continues to grow, there has been a corresponding development in ecological marketing in response to society's pro-ecological demands. Unlike conventional marketing strategies that often prioritize product or market expansion, the emerging new marketing paradigm places greater emphasis on the qualitative aspects of offerings.

The main objective of this chapter will be to outline the marketing strategies used for sustainable products and describe the tools upon which its principles are based. By means of this exploration, we aim to provide insights into how businesses can effectively engage with consumers who prioritize sustainability, while also meeting their own business objectives.

10.1   Concept of sustainable marketing

Sustainability is attracting more and more attention in various fields, including marketing. The concept of sustainable marketing has evolved through successive

DOI: 10.4324/9781032710693-11

periods, which can be divided chronologically into three periods: ecological marketing in the 1970s, environmental/ecological marketing in the 1980s and mid-1990s, and sustainability marketing since the second half of the 1990s (Leonidou & Leonidou, 2011; Lloveras et al., 2022; Sołtysik et al., 2024). The growing importance of sustainable marketing has also been highlighted in academic research (Harrison et al., 2023; Žabkar et al., 2018), public policy and corporate practices (Iyer & Reczek, 2017).

Saleem et al. (2021) conducted a bibliometric analysis of articles published between 1977 and 2020 in the Web of Science (WOS) database with the aim of summarizing the current state of green marketing research as well as presenting and analysing the search results using selected keywords ("Green marketing" OR "ecological marketing" OR "eco* marketing" OR "sustainable marketing" OR "Environmental Marketing" OR "Enviro* Marketing"). The study revealed a growing interest among researchers in the subject of green marketing. The number of published articles has steadily increased over the years, especially in the last decade. The largest number of articles appeared in 2019. This upward trajectory is in line with the current global trend in both corporate social responsibility (CSR) and sustainability research (Saleem et al., 2021). Also, Kar and Harichandan (2022) focused on the Scopus and WOS databases containing data from 1,121 articles published between 1990 and 2021 in 462 journals. The results suggest that green marketing strategies and sustainable marketing techniques are growing in importance (Kumar Kar & Harichandan, 2022).

It should be pointed out that sustainable marketing is a term commonly associated with sustainability. However, sustainability as a concept can be controversial. It is open to many interpretations and difficult to translate into meaningful actions due to the numerous political, economic, and technological constraints that companies and governments face in this area (Ruggiero et al., 2021). Sustainable marketing undoubtedly has a strong axiological dimension, as does the idea of sustainable development, and the overriding benefit of both concepts is a better quality of life for current and future generations (Sołtysik et al., 2024). Both sustainable development and sustainable marketing cover three main areas, that is, environmental, social, and economic, while green marketing, for example, focuses mainly on environmental factors. Research by Bhardwaj et al. (2023) indicates that green marketing and sustainability are increasingly focused on social aspects, suggesting that researchers are nowadays placing more emphasis on the marketing process, thereby stimulating green marketing (Bhardwaj et al., 2023). Caring for the environment is becoming more and more important for consumers, both in terms of products and production processes. Therefore, green marketing practices and efforts are expanding to encompass processes, goods, and services (Polonsky, 2011).

One distinctive form of social marketing is ecologically responsible marketing, which is a specific reaction to the pro-ecological needs of consumers and the difficulties associated with a pro-ecological lifestyle that arise from a conflict between the desire to meet individual and current needs, on the one hand, and the long-term interests of consumers, on the other (Roberts, 1995). One of the main goals of organic marketing is to resolve this conflict. Therefore, it is not surprising that the development of ecological marketing is, in a sense, a consequence of the growing environmental awareness of consumers, which itself is a response to the pro-ecological needs of society (Ellen, 1994). Nowadays, marketing is more and more concentrated on the buyer, which means that marketing is not only about sales and advertising but also covers a wide range of additional activities, such as product design and redesign, green production, and integrated marketing communication. In conventional marketing, the dominant strategy has been the philosophy of product or market expansion, as opposed to new marketing, which for the most part emphasizes the qualitative characteristics of products and services. Thus, the main strategic objectives of a company in sustainable marketing include (Park et al., 2022a) the following:

- the image of an ethical company, and thus an honest one, representing certain values and offering certain benefits,
- survival and growth through strong customer loyalty,
- social and ecological marketing, and
- financial goals based on a reasonable compromise between the needs of the company and the needs of customers, thus ensuring growth.

However, there is still no clear definition of sustainable marketing in the marketing literature (Lunde, 2018). The Brundtland Commission's definition of sustainability and Elkington's concept of the Triple Bottom Line have one major weakness – they fail to consider the exchange of value that is a key element of marketing (Park et al., 2022a). This, in turn, has important implications for marketing strategies and research on consumer behaviour. To address this issue, Lunde (2018) proposes a definition of sustainable marketing as the strategic process of creating, communicating, delivering, and exchanging products and services that generate value through consumer behaviour, business practices, and in the marketplace, while minimizing environmental damage and improving the quality of life and the well-being of consumers and global stakeholders in an ethical and equitable manner, both today and for future generations (Lunde, 2018). Sustainable marketing also encompasses socially or environmentally conscious products and services – beneficial to the planet and society as a whole.

In view of the above analysis, we can conclude that the key feature of sustainable marketing is, first of all, the adoption of a systemic approach and the revision of existing paradigms. However, this is not synonymous with a rejection of the achievements of traditional marketing, but rather a shift in attention from purely economic criteria to a perspective that also embraces socio-ecological aspects. This means that marketing activities that take into account environmental issues are becoming an integral part of an organization's strategy, and not just an additional element or a "fashionable" trend (Menon & Menon, 1997). In this way, green marketing constitutes a more holistic approach that takes heed not only of financial gains but also of an organization's impact on society and the environment. Thus, sustainable marketing can be considered in three aspects (Žabkar et al., 2018):

1 In a conceptual dimension, based on the philosophy of management in a modern market organization, in which attention is focused on customers and their growing need take care of the environment. In this view, the customer becomes the focal point of interest, and competition is focused on meeting these needs while being aware of environmental protection.
2 In a decision-making dimension, where the concept of eco-marketing must be operationalized by establishing formal organizational structures that enable decisions on sustainable marketing issues.
3 In an instrumental dimension, where the main tools used in marketing are those that enable companies to acquire new customers, retain existing ones, and compete on the market to strengthen their position.

From the point of view of achieving set objectives, choosing the right marketing instruments and the optimal ways of performing marketing tasks is of great importance to enterprises. Through a properly planned and implemented marketing policy, an organization can both perform tasks and do so in a way that ensures the most favourable relationship between inputs and effects. The changing conditions in which companies operate (both external conditions, e.g., environmental regulations and demand for specific products, and internal, e.g., investment opportunities, lower costs, and increasing potential) are compelling companies to adopt specific behaviours towards environmental problems. For example, according to Kotler (2011), environmentalists are seeking to bring about a change in marketing practices, and at the core of green marketing (strategic and functional) lies a search for opportunities and decisions that can be exploited by adopting predominantly green marketing practices (Kotler, 2011). Instead, Ivanović-Đukić et al. (2020) focus on factors shaping the development of sustainable entrepreneurship, such as financial incentives, measures designed to facilitate market access, technical assistance in the development of green products, advice in their marketing and distribution, and measures encouraging sustainable entrepreneurship, such as sustainable incubators (Ivanović-Đukić et al., 2020).

However, if we look at the methodology and norms, the two areas that have the greatest impact on the field of marketing and consumer behaviour are economics and psychology. Marketers, drawing on economics and psychology, regard the individual/consumer as the primary unit of analysis, and have adopted a quantitative, positivist approach to research (Davies et al., 2020). The contribution of psychology to the field of marketing is extremely important, especially in the context of sustainable consumption and marketing (Piligrimienė et al., 2021). While the impact of economics on sustainability research remains significant, there has long been a debate over the rationality of the assumption that individuals are rational, analytical decision-makers (Bagozzi, 1975). Sustainability research in marketing is dominated by theories such as planned behaviour theory and ethical decision-making, which take into account values, beliefs, and norms, but without demonstrating that attitudes have a significant impact on the behaviour of individuals (Kalafatis et al., 1999). Sometimes, a change in behaviour can occur without a change in attitudes or intentions (Davies et al., 2020). In addition, if we look at the social role of sustainable consumer behaviour, the main motivators of this behaviour are social norms, social influence, culture, marketing activities, and social media (Xiao et al., 2023).

10.2 Tools supporting sustainable marketing

Sustainable marketing tools comprise those actions and measures employed in the practice of sustainable marketing that aim to shape consumer and business behaviour in such a way that it benefits both consumers and the social environment. Excessive consumerism clearly has a negative impact on both the environment and the well-being of the individual (Akenji, 2014). Therefore, responsible environmental behaviour not only involves reducing consumption and adopting a sustainable lifestyle but also adopting anti-consumption measures and living according to one's own abilities (Ziesemer et al., 2021). Concern for the environment, previous experience and perceived utility have a significant impact on the ecological behaviour of consumers. On the other hand, a lack of knowledge about environmental issues, high prices and associated risks are barriers preventing positive attitudes being converted into actual purchases of green products (Sharma, 2021; Sołtysik et al., 2024).

Many companies want to give the impression that they are achieving sustainability by using methods that are often only designed to increase profits without actually promoting business sustainability (Roscoe et al., 2019). In addition, the demand generated by traditional advertising is based on a "push" rather than a "pull" marketing strategy (Unni & Harmon, 2007). This may induce customers to buy products in the short term, but it does not lead to an internal transformation of consumer values into those consistent with the company's sustainable values (Park et al., 2022b). As a result, when it comes to both the actions taken and the tools used in sustainable marketing, there needs to be a shift away from aggressive sales techniques towards a greater focus on long-term strategies that aim to prove if there is any real value to environmental claims. For exchange participants to achieve satisfaction in accordance with the principles of sustainable development, the primacy of common and long-term needs connected with sustainable development should be given primacy over the individual and short-term needs of the exchange parties. Both suppliers and customers should consider the Sustainable Development Goals (SDGs) as the main guidelines for meeting other needs. In this approach, the environment is also treated as a party to the exchange, since each exchange can have a greater or lesser impact on nature as the place from which raw materials are taken as well as on nature as the place in which substances are disposed (Fallah Shayan et al., 2022).

Therefore, when using sustainable marketing tools, companies today need to gain insights into the determinants of responsible behaviour, because consumers are a key stakeholder in the success of companies that implement socially responsible strategies. Research also shows that social and personal norms, sustainability concerns, and consumer awareness influence sustainable behaviour (Hosta & Zabkar, 2021; Sołtysik et al., 2024). For example, social norms have a significant impact on choices regarding green consumption, energy conservation, recycling, and water conservation (Cialdini & Jacobson, 2021). Managers can thus promote sustainable environmental and social behaviour, thereby promoting pro-environmental standards. Highlighting the benefits of sustainable behaviour to increase consumer satisfaction may encourage more sustainable consumption, especially among those consumers initially unconvinced by green consumer practices (Ramos-Hidalgo et al., 2022).

Given the belief of many consumers that companies should demonstrate a commitment to social responsibility, it is important to understand how CSR activities affect how willing customers are to choose the products offered by these companies. Research conducted as part of a systematic literature review by Narayanan and Singh confirms the overall positive impact of CSR on customers' desire to behave in such a way. These findings provide vital guidance for marketers, helping them develop pricing strategies based on a better understanding of consumer behaviour related to CSR perceptions (Narayanan & Singh, 2023).

Despite companies' efforts, consumers still perceive green or environmentally friendly products and services as more expensive and less effective compared to conventional products (Kabaja et al., 2022). This perception may be due to the fact that sustainable marketing strategies often focus mainly on social and environmental issues, while ignoring those connected with cost, utility, and quality. The traditional marketing mix, which includes product, price, venue, and promotion, still needs to be incorporated into a marketing strategy that emphasizes the importance of providing solutions for customers, controlling costs, providing convenience, and effective communication (Wongleedee, 2015). Integrating sustainable marketing with customer value generation may give rise to business practices based on a more sustainable approach. As a result, companies are increasingly under pressure to focus on sustainable business development and achieving sustainable value creation while generating profits (Evans et al., 2017).

Key CSR practices include responsible sourcing, responsible supplier selection and management, and the strategic integration of CSR, which together should help generate triple results (social, environmental, and economic) (Lo, 2020). Specific tools for sustainable marketing can be found in the ways CSR is communicated to stakeholders (Gosselt et al., 2019). The main forms of sustainable communication include not only sustainable/green advertising but also sustainability reporting.

Investing corporate resources in addressing social, environmental, and governance issues can deliver shareholder value and improve the bottom line, which benefits the company and its stakeholders. The better a company's environmental, social and governance performance, the greater the impact of CSR on this bottom line (Coelho et al., 2023).

With the increasing stress on social and environmental responsibility, companies are faced with a need to balance relationships with customers, while striving to augment their business value as well as focus on customer needs. This requires taking into account social, economic, and environmental impacts and at the same time engaging with sustainability-conscious customers in order to build lasting relationships with them. In marketing strategies based on CSR, it is important to be aware of customer perceptions, as how the latter gauge the compliance of sustainable strategies with their own expectations has a significant impact on the results of companies (Gleim et al., 2023).

Since consumers take into account many factors – economic, environmental and social – when choosing sustainable products we can conclude that the decision-making process is more complex and involves more factors than is the case with products lacking sustainability features. To help customers make informed choices, companies use eco-labels. Because these offer independent verification of environmental safety, they are more trusted than manufacturers' own claims of quality (Wojnarowska et al., 2021).

Although the conceptual framework for the Ecolabel scheme was first initiated by Germany and Scandinavia in 1977 (Browner, 1998; Mufidah et al., 2018), semiotics as the study of signs, their nature and role in the process of cognition was discussed by J. Locke back in 1690. The latter argued that the sign is an intermediary between an object examined by reason and reason itself. In the case of eco-labelling, a correct interpretation of a label by means of rapid perception (primarily visual) relies on knowledge resources associated with the eco-label (Sammer & Wüstenhagen, 2006) and can help consumers identify an environmentally friendly product and decide to purchase it. Mitigating a negative impact on the environment is undoubtedly one benefit of choosing organic products and thus helps achieve the objectives of sustainable production and sustainable consumption (Wojnarowska et al., 2021).

One of the key objectives of the ecolabel is to provide reliable and credible information on the environmental impact of goods, making it easier for consumers to choose environmentally friendly products.

Today, there are many different types and varieties of eco-labels and certification that provide information about the sustainability of products (Minkov et al., 2018). In recent decades, eco-labelling has become a strategic tool of communication for environmentally friendly products (Bougherara & Combris, 2009; Clemenz, 2010; Song et al., 2019). To this end, companies use eco-labels, which inform consumers about how their products are less harmful to the environment (Sistemi et al., 2008). In addition to making long-term profits, companies use eco-labels to differentiate themselves from the competition and demonstrate greater CSR (Ban & Iacobaş, 2016; Yilmaz et al., 2019). According to Gallastegui, the ecolabel has two primary purposes. The first objective is to inform consumers about the environmental impact of consumption and to encourage consumers to change their attitudes towards sustainable consumption patterns. The second objective is to encourage governments, manufacturers, and other suppliers to create services and goods in accordance with environmental standards (Gallastegui, 2002). In this regard, it can be concluded that another objective of eco-labelling is to create a demand for goods that are more desirable from an environmental point of view and, consequently, to induce producers to supply goods that meet these expectations. In this context, the eco-label is seen as one of the best tools for promoting organic products, influencing consumers' purchasing decisions (Aertsens et al., 2011; Smith & Paladino, 2010; Yau, 2012).

The eco-label has a predetermined graphic form and constitutes proof that the manufacturer meets certain standards. Hence, organic products cannot be labelled at their own discretion, even though the placement of eco-labels on products is done voluntarily. The requirements that an ecolabel must meet are determined by the Global Ecolabelling Network (GEN, 2004).

If consumers understand the importance of ecological benefits of certain features, they will be more willing to pay an ecological premium (Teisl et al., 2002). As a consequence, eco-labelling is an important tool for increasing consumer confidence in environmentally friendly products and services (Song et al., 2019). Consumer attitudes towards eco-labels matter because consumers should be aware of the meaning communicated by different eco-labels and trust them. (Buunk & van der Werf, 2019; Gössling & Buckley, 2016). On the other hand, both deficiencies in supervision and an imperfect regulatory system have also created opportunities for eco-labels to be misused on the market (Lyon & Montgomery, 2015), thereby undermining the credibility of the system on the whole (D'Souza et al., 2006).

Research shows that the ethical attributes of products can shape consumer attitudes and are usually an effective indicator of whether positive attitudes are turning into actual purchasing behaviour (Young et al., 2010). Research also confirms d that the presence of the word "green" has a significant impact on sales volume (Cerjak et al., 2010). Simon (1992) suggests that products labelled as environmentally friendly reflect new attitudes towards environmental values. Environmental awareness is considered a prerequisite for drawing consumers' attention to the essential features of a product (Thøgersen et al., 2010).

10.3   The future of sustainable product marketing

In the evolving landscape of consumer markets, the future of sustainable product marketing is poised at a pivotal juncture, driven by the steadily increasing ecological awareness of consumers and the profound impact of digital technologies.

Organizations need to be cognizant of this paradigm shift. In an era when consumer preferences are markedly influenced by environmental and ethical considerations (De Medeiros, Ribeiro & Cortimiglia, 2014), sustainable product marketing is emerging as an increasingly key factor. Research in the field of marketing has shown that modern consumers are being drawn more and more to eco-friendly products, which shows that sustainable products marketing will continue to perform an ever more valuable role (Averdung & Wagenfuehrer, 2011).

A study conducted by Hanss and Böhm (2011) shows that all dimensions of sustainability are vital in shaping consumers' understanding of this concept, namely the environmental, social and economic dimensions. As regards those attributes of sustainable products deemed important, consumers placed considerable emphasis on recyclable packaging, fair payments from producers, low energy consumption and low carbon dioxide emissions during production and shipping (Hanss & Böhm, 2011; Chirilli, Molino & Torri, 2022). In the context of sustainable marketing products, a growing body of research indicates a significant shift in consumer awareness of sustainability. Building on these insights, the future of sustainable product marketing should focus on amplifying these key attributes – recyclability, fair trade practices, energy efficiency, and reduced carbon emissions – as they align closely with the evolving preferences and ethical considerations of consumers. This strategic emphasis can significantly enhance the appeal and market success of sustainable products.

This heightened awareness is not simply a fleeting trend but rather a profound change in consumer behaviour, one that is deeply rooted in a greater understanding of the sustainable development context (Aprile & Punzo, 2022). As this awareness continues to increase, it is expected to have an increasingly substantial effect on purchasing decisions (García-Salirrosas, Rondon-Eusebio, 2022; Siraj et al., 2022). This evolving dynamic presents a pivotal opportunity for the future of sustainable product marketing. Brands and companies that effectively integrate and highlight sustainable practices and values within their products and services are likely to find favour with a more environmentally and socially conscious consumer base (Cohen & Muñoz, 2017).

The relentless advances in digital technology are enabling more accurate tracking of the environmental and societal impact of products, their supply chains, and life cycles. In particular, the emergence of blockchain offers a unique lens through which to scrutinize and predict its transformative effects on sustainable products marketing. It equips consumers with powerful tools to verify the authenticity of sustainable claims, thereby mitigating the prevalence of greenwashing. Blockchain technology serves as a pivotal link between eco-labels and industry initiatives, offering a clear view of the transparency of sustainable business practices (Navas et al., 2021). The inherent transparency, reliability, and unalterable record-keeping features of blockchain have the potential to transform industries with unsustainable practices, fostering social impact (Lapointe & Fishbane, 2019) and it has the potential to revolutionize sustainable product marketing by providing business practices and eco-labels with a new level of transparency and credibility. By embedding this technology within the retail sector, it empowers consumers, thereby strengthening and encouraging their participation in sustainable consumption behaviour

(Hartmann, Apaolaza & D'Souza, 2018). Such a situation can enable consumers to accurately trace the history and origin of products and thus also their true sustainability. Consequently, the use of blockchain in sustainable product marketing not only increases consumer confidence but also stimulates the development of more sustainable practices throughout the supply chain. This, in turn, could lead to a deeper shift in consumer behaviour and accelerate the transition to more sustainable business models (Calandra et al., 2023).

One possible future trend may be the use of state-of-the-art technologies such as, for instance, neuromarketing techniques and their applications in sustainable product marketing. Research conducted by Nilashi et al. (2020) indicates that neuromarketing techniques can help ensure a brand keeps the attention of consumers more effectively. Moreover, they can also be used on a more intensive basis to develop the most successful branding strategies for green products.

Artificial intelligence (AI) is already having a significant impact on organizations as well as on various aspects of management (Jarosz, 2023; Sołtysik & Jarosz, 2023). The burgeoning role of AI is shaping marketing strategies, with its predictive and analytical capabilities capable of refining consumer targeting and enhancing the transparency of sustainability claims. Herman (2021) stated that AI has the following benefits:

- it can guide the design and development of products and services by foreseeing the sustainable features of products/services most valued by consumers,
- AI-enabled income prediction based on digital footprints can personalize prices according to consumers' potential willingness to pay for sustainable products, and
- it can effectively match sustainable products with consumer groups that are most suited for certain types of products and services (for instance, in terms of place and promotion).

In the light of the above considerations, it is important to stress that one factor that will be pivotal to the future trajectory of sustainable product marketing will be the critical interplay between technological advances, increasing consumer awareness, and evolving market dynamics in shaping a more sustainable and ethically conscious marketplace.

In addition, it is clear that the future of sustainable product marketing will also be determined and shaped by the growing share of such products in companies' product portfolios since sustainability is becoming more common and companies around the world are adopting SDGs (Howard-Grenville et al., 2019; Erin, Bamigboye & Oyewo, 2022).

10.4   Conclusions

Many companies have made significant investments in sustainable marketing. Unilever is striving to establish its credibility and corporate identity as a sustainable business by setting sustainability goals in its production and supply chain management. Panasonic produces eco-friendly products, thereby contributing to sustainable development in daily business life. Moreover, it is collaborating with many companies in the creation of a sustainable smart city exhibition in Japan. Allergan, a California-based pharmaceutical company, has been implementing the concept of sustainability in its business processes for more than 20 years, focusing on the areas of water management, environmental protection, energy management, waste and emissions reduction in business operations, and supply chain management. Patagonia encourages its consumers to buy only the essentials and has put in place a program to repair existing products instead of replacing them with new ones. The company's commitment to sustainability is evident in its products and messaging. As early as the 1960s, IBM adopted the principles of sustainable development and nowadays is focusing on the construction of smart buildings, reducing demand for resources, promoting green procurement, ensuring efficient management water resources, and adopting a broad-based approach (Park et al., 2022b).

One challenge facing modern marketing is that of integrating the principles of sustainable development into its strategies and practices in response to the growing environmental awareness of consumers and the introduction of laws and regulations in this area. Sustainable marketing as a strategy encompassing corporate activities aimed at integrating ecological, social, and economic goals is a response to the growing demands of both consumers and regulators. Examples from companies such as Unilever, Panasonic, Allergan, Patagonia, and IBM highlight the diverse approaches to implementing the principles of sustainability, ranging from supply chain management and promoting green products to initiatives designed to reduce environmental impact. These activities not only have a positive impact on the company's image, but also help generate long-term benefits for both shareholders and society.

At the same time, the new regulations introduced by the European Parliament are aimed at combating greenwashing, that is, unfair marketing practices whereby companies attribute ecological characteristics to products without real justification. Amendments to Directives 2005/29/EC and 2011/83/EU strengthen consumer protection by introducing stricter requirements for environmental claims and product labelling. They prohibit, among other things, making environmental claims without clear commitments, affixing labelling not based on any certification scheme, advertising negligible benefits and suggesting the superiority of products without any objective grounds for doing so.

This clash between private initiatives and regulatory requirements shows how rapidly the field of sustainable marketing is developing, and highlighting the need for transparency, authenticity, and responsibility in how a company communicates its brand to consumers. In the face of these challenges, companies must not only adapt their activities and marketing strategies to the new regulatory framework, but also meet societal expectations regarding sustainable development.

Acknowledgement

The publication/article presents the result of the Project no 070/ZJE/2024/POT financed from the subsidy granted to the Krakow University of Economics.

References

Aertsens, J., Mondelaers, K., Verbeke, W., Buysse, J., & Van Huylenbroeck, G. (2011). The influence of subjective and objective knowledge on attitude, motivations and consumption of organic food. *British Food Journal, 113*(11), 1353–1378. https://doi. org/10.1108/00070701111179988

Akenji, L. (2014). Consumer scapegoatism and limits to green consumerism. *Journal of Cleaner Production, 63*, 13–23. https://doi.org/10.1016/j.jclepro.2013.05.022

Aprile, M. C., & Punzo, G. (2022). How environmental sustainability labels affect food choices: Assessing consumer preferences in southern Italy. *Journal of Cleaner Production, 332*, 130046. https://doi.org/10.1016/j.jclepro.2021.130046

Averdung, A., & Wagenfuehrer, D. (2011). Consumers acceptance, adoption and behavioural intentions regarding environmentally sustainable innovations. *Journal of Business Management and Economics, 2*, 98–106.

Ban, O. I., & Iacobaş, P. (2016). Marketing research regarding tourism business readiness for eco-label achievement (case study: Natura 2000 crişul repede gorge-pădurea craiului pass site, Romania). *Ecoforum, 5*(1), 224–234.

Bhardwaj, S., Nair, K., Tariq, M. U., Ahmad, A., & Chitnis, A. (2023). The state of research in green marketing: A bibliometric review from 2005 to 2022. *Sustainability, 15*(4), 2988. https://doi.org/10.3390/su15042988

Bougherara, D., & Combris, P. (2009). Eco-labelled food products: What are consumers paying for? *European Review of Agricultural Economics, 36*(3), 321–341. https://doi. org/10.1093/erae/jbp023

Browner, C. M. (1998). Guidelines for Ecological Risk Assessment. Risk Assessment Forum U.S. Environmental Protection Agency Washington, DC, April 2018

Buunk, E., & van der Werf, E. (2019). Adopters versus non-adopters of the Green Key eco-label in the Dutch accommodation sector. *Sustainability (Switzerland), 11*(13), 187–192. https://doi.org/10.3390/su11133563

Calandra, D., Secinaro, S., Massaro, M., Dal Mas, F., & Bagnoli, C. (2023). The link between sustainable business models and Blockchain: A multiple case study approach. *Business Strategy and the Environment, 32*(4), 1403–1417.

Cerjak, M., Mesić, Ž., Kopić, M., Kovačić, D., & Markovina, J. (2010). What motivates consumers to buy organic food: Comparison of Croatia, Bosnia Herzegovina, and Slovenia. *Journal of Food Products Marketing, 16*(3), 278–292. https://doi.org/10.1080/ 10454446.2010.484745

Chirilli, C., Molino, M., & Torri, L. (2022). Consumers' awareness, behavior and expectations for food packaging environmental sustainability: Influence of socio-demographic characteristics. *Foods, 11*(2388). https://doi.org/10.3390/foods11162388

Cialdini, R. B., & Jacobson, R. P. (2021). Influences of social norms on climate change-related behaviors. *Current Opinion in Behavioral Sciences, 42*, 1–8. https://doi.org/10.1016/j. cobeha.2021.01.005

Clemenz, G. (2010). Eco-labeling and horizontal product differentiation. *Environmental and Resource Economics, 45*(4), 481–497. https://doi.org/10.1007/s10640-009-9324-2

Coelho, R., Jayantilal, S., & Ferreira, J. J. (2023). The impact of social responsibility on corporate financial performance: A systematic literature review. *Corporate Social Responsibility and Environmental Management, 30*(4), 1535–1560. https://doi.org/10.1002/csr.2446

Cohen, B., & Muñoz, P. (2017). Entering conscious consumer markets: Toward a new generation of sustainability strategies. *California Management Review, 59*(4), 23–48. https://doi.org/10.1177/0008125617722792

D'Souza, C., Taghian, M., Lamb, P., & Peretiatkos, R. (2006). Green products and corporate strategy: An empirical investigation. *Society and Business Review, 1*(2), 144–157. https://doi.org/10.1108/17465680610669825

Davies, I., Oates, C. J., Tynan, C., Carrigan, M., Casey, K., Heath, T., Henninger, C. E., Lichrou, M., McDonagh, P., McDonald, S., McKechnie, S., McLeay, F., O'Malley, L., & Wells, V. (2020). Seeking sustainable futures in marketing and consumer research. *European Journal of Marketing, 54*(11), 2911–2939. https://doi.org/10.1108/EJM-02-2019-0144

De Medeiros, J. F., Ribeiro, J. L. D., & Cortimiglia, M. N. (2014). Success factors for environmentally sustainable product innovation: A systematic literature review. *Journal of Cleaner Production, 65*, 76–86.

Ellen, P. S. (1994). Do we know what we need to know? Objective and subjective knowledge effects on pro-ecological behaviors. *Journal of Business Research, 30*(1), 43–52. https://doi.org/10.1016/0148-2963(94)90067-1

Erin, O. A., Bamigboye, O. A., & Oyewo, B. (2022). Sustainable development goals (SDG) reporting: An analysis of disclosure. *Journal of Accounting in Emerging Economies, 12*(5), 761–789. https://doi.org/10.1108/JAEE-02-2020-0037

Evans, S., Vladimirova, D., Holgado, M., Van Fossen, K., Yang, M., Silva, E. A., & Barlow, C. Y. (2017). Business model innovation for sustainability: Towards a unified perspective for creation of sustainable business models. *Business Strategy and the Environment, 26*(5), 597–608. https://doi.org/10.1002/bse.1939

Fallah Shayan, N., Mohabbati-Kalejahi, N., Alavi, S., & Zahed, M. A. (2022). Sustainable Development Goals (SDGs) as a framework for Corporate Social Responsibility (CSR). *Sustainability, 14*(3), 1222. https://doi.org/10.3390/su14031222

Gallastegui, I. G. (2002). The use of eco-labels: A. *331*, 316–331. https://doi.org/10.1002/eet.304

García-Salirrosas, E. E., & Rondon-Eusebio, R. F. (2022). Green marketing practices related to key variables of consumer purchasing behavior. *Sustainability, 14*(14), 8499. https://doi.org/10.3390/su14148499

Gleim, M. R., McCullough, H., Sreen, N., & Pant, L. G. (2023). Is doing right all that matters in sustainability marketing? The role of fit in sustainable marketing strategies. *Journal of Retailing and Consumer Services, 70*, 103124. https://doi.org/10.1016/j.jretconser.2022.103124

Gosselt, J. F., van Rompay, T., & Haske, L. (2019). Won't get fooled again: The effects of internal and external CSR ECO-labeling. *Journal of Business Ethics, 155*(2), 413–424. https://doi.org/10.1007/s10551-017-3512-8

Gössling, S., & Buckley, R. (2016). Carbon labels in tourism: Persuasive communication? *Journal of Cleaner Production, 111*, 358–369. https://doi.org/10.1016/j.jclepro.2014.08.067

Hanss, D., & Böhm, G. (2011). Sustainability seen from the perspective of consumers. *International Journal of Consumer Studies, 36*(6), 678–687. https://doi.org/10.1111/j.1470-6431.2011.01045.x

Harrison, D., Prenkert, F., Hasche, N., & Carlborg, P. (2023). Business networks and sustainability: Past, present and future. *Industrial Marketing Management, 111*, A10–A17. https://doi.org/10.1016/j.indmarman.2023.03.011

Hartmann, P., Apaolaza, V., & D'Souza, C. (2018). The role of psychological empowerment in climate-protective consumer behavior. *European Journal of Marketing, 52*(3/4), 392–417.

Hermann, E. (2023). Artificial intelligence in marketing: Friend or foe of sustainable consumption? *AI & Society, 38*, 1975–1976. https://doi.org/10.1007/s00146-021-01227-8

Hosta, M., & Zabkar, V. (2021). Antecedents of environmentally and socially responsible sustainable consumer behavior. *Journal of Business Ethics, 171*(2), 273–293. https://doi.org/10.1007/s10551-019-04416-0

Howard-Grenville, J., Davis, G. F., Dyllick, T., Miller, C. C., Thau, S., & Tsui, A. S. (2019). Sustainable development for a better world: Contributions of leadership, management, and organizations. *Academy of Management Discoveries, 5*(4). https://doi.org/10.5465/amd.2019.0275

Ivanović-Đukić, M., Petrović-Ranđelović, M., & Talić, M. (2020). An analysis of factors influencing the development of social enterprises in the Republic of Serbia. *The European Journal of Applied Economics, 17*(2), 1–18. https://doi.org/10.5937/EJAE17-27375

Iyer, E. S., & Reczek, R. W. (2017). The intersection of sustainability, marketing, and public policy: Introduction to the special section on sustainability. *Journal of Public Policy & Marketing, 36*(2), 246–254. https://doi.org/10.1509/jppm.36.250

Jarosz, S. (2023). Artificial intelligence – An agenda for management sciences. *e-mentor, 2*(99), 47–55. https://doi.org/10.15219/em99.1603

Kabaja, B., Wojnarowska, M., Cesarani, M. C., & Varese, E. (2022). Recognizability of ecolabels on e-commerce websites: The case for younger consumers in Poland. *Sustainability, 14*(9), 5351. https://doi.org/10.3390/su14095351

Kalafatis, S. P., Pollard, M., East, R., & Tsogas, M. H. (1999). Green marketing and Ajzen's theory of planned behaviour: A cross-market examination. *Journal of Consumer Marketing, 16*(5), 441–460. https://doi.org/10.1108/07363769910289550

Kotler, P. (2011). Reinventing marketing to manage the environmental imperative. *Journal of Marketing, 75*(4), 132–135. https://doi.org/10.1509/jmkg.75.4.132

Kumar Kar, S., & Harichandan, S. (2022). Green marketing innovation and sustainable consumption: A bibliometric analysis. *Journal of Cleaner Production, 361*, 132290. https://doi.org/10.1016/j.jclepro.2022.132290

Lapointe, C., & Fishbane, L. (2019). The blockchain ethical design framework. *Innovation in Technology, Governance, Globalization, 12*, 50–71.

Leonidou, C. N., & Leonidou, L. C. (2011). Research into environmental marketing/management: A bibliographic analysis. *European Journal of Marketing, 45*(1/2), 68–103. https://doi.org/10.1108/03090561111095603

Lloveras, J., Marshall, A. P., Vandeventer, J. S., & Pansera, M. (2022). Sustainability marketing beyond sustainable development: Towards a degrowth agenda. *Journal of Marketing Management, 38*(17–18), 2055–2077. https://doi.org/10.1080/0267257X.2022.2084443

Lo, A. (2020). Effects of customer experience in engaging in hotels' CSR activities on brand relationship quality and behavioural intention. *Journal of Travel and Tourism Marketing, 37*(2). https://doi.org/10.1080/10548408.2020.1740140

Lunde, M. B. (2018). Sustainability in marketing: A systematic review unifying 20 years of theoretical and substantive contributions (1997–2016). *AMS Review, 8*(3–4), 85–110. https://doi.org/10.1007/s13162-018-0124-0

Lyon, T. P., & Montgomery, A. W. (2015). The means and end of greenwash. *Organization & Environment, 28*(2), 223–249. https://doi.org/10.1177/1086026615575332

Menon, A., & Menon, A. (1997). Enviropreneurial marketing strategy: The emergence of corporate environmentalism as market strategy. *Journal of Marketing, 61*(1), 51–67. https://doi.org/10.1177/002224299706100105

Minkov, N., Bach, V., & Finkbeiner, M. (2018). Characterization of the Cradle to Cradle CertifiedTM products program in the context of eco-labels and environmental declarations. *Sustainability (Switzerland), 10*(3). https://doi.org/10.3390/su10030738

Mufidah, I., Jiang, B. C., Lin, S. C., Chin, J., Rachmaniati, Y. P., & Persada, S. F. (2018). Understanding the consumers' behavior intention in using green ecolabel product through Pro-Environmental Planned Behavior model in developing and developed regions: Lessons learned from Taiwan and Indonesia. *Sustainability (Switzerland), 10*(5), 1–15. https://doi.org/10.3390/su10051423

Narayanan, S., & Singh, G. A. (2023). Consumers' willingness to pay for corporate social responsibility: Theory and evidence. *International Journal of Consumer Studies, 47*(6), 2212–2244. https://doi.org/10.1111/ijcs.12910

Navas, R., Chang, H. J., Khan, S., & Chong, J. W. (2021). Sustainability transparency and trustworthiness of traditional and blockchain ecolabels: A comparison of generations X and Y consumers. *Sustainability, 13*(15), 8469. https://doi.org/10.3390/su13158469

Network, Global Ecolabelling Network (GEN) (2004). Information Paper: Introduction to Ecolabelling Prepared July 2004 Introduction to Ecolabelling. *July*, https://www.gdrc.org/sustbiz/green/gen-infopaper.pdf

Nilashi, M., Yadegaridehkordi, E., Samad, S., Mardani, A., Ahani, A., Aljojo, N., Razali, N. S., & Tajuddin, T. (2020). Decision to adopt neuromarketing techniques for sustainable product marketing: A fuzzy decision-making approach. *Symmetry, 12*(2), 305. https://doi.org/10.3390/sym12020305

Park, J. Y., Perumal, S. V., Sanyal, S., Ah Nguyen, B., Ray, S., Krishnan, R., Narasimhaiah, R., & Thangam, D. (2022a). Sustainable marketing strategies as an essential tool of business. *The American Journal of Economics and Sociology, 81*(2), 359–379. https://doi.org/10.1111/ajes.12459

Park, J. Y., Perumal, S. V., Sanyal, S., Ah Nguyen, B., Ray, S., Krishnan, R., Narasimhaiah, R., & Thangam, D. (2022b). Sustainable marketing strategies as an essential tool of business. *The American Journal of Economics and Sociology, 81*(2), 359–379. https://doi.org/10.1111/ajes.12459

Piligrimienė, Ž., Banytė, J., Dovalienė, A., Gadeikienė, A., & Korzilius, H. (2021). Sustainable consumption patterns in different settings. *Engineering Economics, 32*(3), 278–291. https://doi.org/10.5755/j01.ee.32.3.28621

Polonsky, M. J. (2011). Transformative green marketing: Impediments and opportunities. *Journal of Business Research, 64*(12), 1311–1319. https://doi.org/10.1016/j.jbusres.2011.01.016

Ramos-Hidalgo, E., Diaz-Carrion, R., & Rodríguez-Rad, C. (2022). Does sustainable consumption make consumers happy? *International Journal of Market Research, 64*(2), 227–248. https://doi.org/10.1177/14707853211030482

Roberts, J. A. (1995). Profiling levels of socially responsible consumer behavior: A cluster analytic approach and its implications for marketing. *Journal of Marketing Theory and Practice, 3*(4), 97–117. https://doi.org/10.1080/10696679.1995.11501709

Roscoe, S., Subramanian, N., Jabbour, C. J. C., & Chong, T. (2019). Green human resource management and the enablers of green organisational culture: Enhancing a firm's environmental performance for sustainable development. *Business Strategy and the Environment, 28*(5), 737–749. https://doi.org/10.1002/bse.2277

Ruggiero, S., Kangas, H.-L., Annala, S., & Lazarevic, D. (2021). Business model innovation in demand response firms: Beyond the niche-regime dichotomy. *Environmental Innovation and Societal Transitions, 39*, 1–17. https://doi.org/10.1016/j.eist.2021.02.002

Saleem, F., Khattak, A., Ur Rehman, S., & Ashiq, M. (2021). Bibliometric analysis of green marketing research from 1977 to 2020. *Publications, 9*(1), 1. https://doi.org/10.3390/publications9010001

Sammer, K., & Wüstenhagen, R. (2006). The influence of eco-labelling on consumer behaviour – Results of a discrete choice analysis for washing machines. *Business Strategy and the Environment, 15*(3), 185–199. https://doi.org/10.1002/bse.522

Simon, F. L. (1992). Marketing green products in the triad. *The Columbia Journal of World Business, 27*, 268–285.

Siraj, A., Taneja, S., Zhu, Y., Jiang, H., Luthra, S., & Kumar, A. (2022). Hey, did you see that label? It's sustainable!: Understanding the role of sustainable labelling in shaping sustainable purchase behaviour for sustainable development. *Business Strategy and the Environment, 31*(7), 2820–2838. https://doi.org/10.1002/bse.3049

Sistemi, Ü., On, P. İ., & Ng, G. M. İ. (2008). Çevre dostu ürün kavramina bütünsel yaklaşim. *Temiz, Electronic Journal of Social Sciences, 26*, 320–333.

Smith, S., & Paladino, A. (2010). Eating clean and green? Investigating consumer motivations towards the purchase of organic food. *Australasian Marketing Journal, 18*(2), 93–104. https://doi.org/10.1016/j.ausmj.2010.01.001

Sołtysik, M., & Jarosz, S. (2023). Trust: A new approach to management. In Sołtysik, M., Gawłowska, M., Sniezynski, B., & Gunia, A. (Eds.), *Artificial Intelligence, Management and Trust* (1st ed.). New York; London: Routledge. 8–18

Sołtysik, M., Wojnarowska, M., Urbaniec, M., Zabkar, V., Ćwiklicki, M., & Varese, E., (2024). *Sustainable Business in the Era of Digital Transformation: Strategic and Entrepreneurial Perspectives*, Abingdon; New York: Routledge.

Song, Y., Qin, Z., & Yuan, Q. (2019). The impact of eco-label on the young Chinese generation: The mediation role of environmental awareness and product attributes in green purchase. *Sustainability (Switzerland), 11*(4). https://doi.org/10.3390/su11040973

Teisl, M. F., Roe, B., & Hicks, R. L. (2002). Can eco-labels tune a market? Evidence from dolphin-safe labeling. *Journal of Environmental Economics and Management, 43*(3), 339–359. https://doi.org/10.1006/jeem.2000.1186

Thøgersen, J., Haugaard, P., & Olesen, A. (2010). Consumer responses to ecolabels. *European Journal of Marketing, 44*(11), 1787–1810. https://doi.org/10.1108/03090561011079882

Unni, R., & Harmon, R. (2007). Perceived effectiveness of push vs. Pull mobile location based advertising. *Journal of Interactive Advertising, 7*(2), 28–40. https://doi.org/10.1080/15252019.2007.10722129

Wojnarowska, M., Sołtysik, M., & Prusak, A. (2021). Impact of eco-labelling on the implementation of sustainable production and consumption. *Environmental Impact Assessment Review, 86*, 106505. https://doi.org/10.1016/j.eiar.2020.106505

Wongleedee, K. (2015). Marketing mix and purchasing behavior for community products at traditional markets. *Procedia – Social and Behavioral Sciences, 197*, 2080–2085. https://doi.org/10.1016/j.sbspro.2015.07.323

Xiao, J., Yang, Z., Li, Z., & Chen, Z. (2023). A review of social roles in green consumer behaviour. *International Journal of Consumer Studies, 47*(6), 2033–2070. https://doi.org/10.1111/ijcs.12865

Yau, Y. (2012). Eco-labels and willingness-to-pay: A Hong Kong study. *Smart and Sustainable Built Environment, 1*(3), 277–290. https://doi.org/10.1108/20466091211287146

Yilmaz, Y., Üngüren, E., & Kaçmaz, Y. Y. (2019). Determination of managers' attitudes towards eco-labeling applied in the context of sustainable tourism and evaluation of the effects of eco-labeling on accommodation enterprises. *Sustainability (Switzerland), 11*(18). https://doi.org/10.3390/su11185069

Young, W., Hwang, K., McDonald, S., & Oates, C. J. (2010). Sustainable consumption: Green consumer behaviour when purchasing products. *Sustainable Development, 18*(1), 20–31. https://doi.org/10.1002/sd.394

Žabkar, V., Kos Koklič, M., McDonald, S., & Abosag, I. (2018). Guest editorial. *European Journal of Marketing, 52*(3/4), 470–475. https://doi.org/10.1108/EJM-04-2018-891

Ziesemer, F., Hüttel, A., & Balderjahn, I. (2021). Young people as drivers or inhibitors of the sustainability movement: The case of anti-consumption. *Journal of Consumer Policy, 44*(3), 427–453. https://doi.org/10.1007/s10603-021-09489-x

11

SUSTAINABLE CONSUMPTION OF PRODUCTS

Marcin Paprocki

11.1 Introduction

From a purely economic perspective, consumerism may be considered a positive phenomenon. Increased consumption of products generally improves economic aspects. Consumerism can contribute to economic growth by generating demand for products and services, while simultaneously stimulating the creation of technological innovations through increased demand for new and better products. On the other hand, excessive consumption can have serious consequences for individuals, society, and the natural environment. In light of the escalating climate crisis, the idea of sustainable consumption is becoming ever more important. In this context, one trend of particular importance is the shift in the consumption model away from consumerism and towards more sustainable consumption patterns. On one hand, it is important that people purchase products that have been sustainably produced, and thus have a positive impact on the economy, the environment, and society. On the other hand, given the negative consequences of consumerism on the environment, efforts should be made to reduce the consumption of products.

11.2 Sustainable consumption of products as a component of sustainable development

When considering consumption in its various guises, emphasis can be placed on one of three factors: meeting needs, the loss of value that of the consumed good and its wear and tear, and the act of purchasing (Iwasiński, 2014). Consumption from the perspective of meeting consumer needs involves satisfying those needs through the consumption of products (goods and services). The consumption of goods encompasses material products such as food and industrial items, while

DOI: 10.4324/9781032710693-12

consumption of services includes both material and non-material services. In the context of environmental and social challenges, the sustainable consumption of products is becoming increasingly important. Sustainable consumption is a complex and ambiguous concept that, due to theoretical fragmentation, is broad in scope but can lead to overlapping ideas (Vargas-Merino et al., 2023). In the 1990s, as the population grew and prosperity increased in societies, consumption rose. As a result, it became increasingly clear that without addressing consumption patterns and levels, achieving the vision of sustainable development might be impossible. As a consequence, the need to address the issue of sustainable consumption became the subject of a major political debate (Mont & Plepys, 2008). The concept of sustainable consumption was defined in Oslo in 1994 as

the use of services and related products, which respond to basic needs and bring a better quality of life while minimizing the use of natural resources and toxic materials as well as the emissions of waste and pollutants over the life cycle of the service or product so as not to jeopardize the needs of future generations.

(Ofstad et al., 1994)

Sustainable consumption assumes the conscious and rational use of available goods and products so as to minimize their harmful effects on the environment and health as well as promote an ecological lifestyle (Ramkissoon et al., 2013). In turn, according to Zalega (2020), sustainable consumption means that individuals consciously strive to minimize the negative effects of the consumption of consumer and investment goods and services by rationalizing and utilising production factors (resources) and reducing the amount of post-production waste and post-consumption waste generated.

Sustainable consumption is one of the fundamental requirements of sustainable development (Wang et al., 2019). There is a consensus among researchers that sustainable development encompasses three dimensions, namely: economic, environmental, and social (Kuhlman & Farrington, 2010; Strezov et al., 2017). The imperative of sustainable development was outlined in the "Report of the World Commission on Environment and Development: Our Common Future". This document defines sustainable development as "meeting the needs of the present without compromising the ability of future generations to meet their own needs" (Brundtland, 1987).

In 2015, the United Nations organized a Sustainable Development Summit in New York. During this summit, 17 Sustainable Development Goals (SDGs) were adopted for the period 2016–2030. These goals were selected as reference points and as a universal roadmap in the transition towards sustainable development under Agenda 2030 for sustainable development (Voulvoulis et al., 2022). One of these reference points is Goal 12 of the SDGs: ensuring sustainable consumption and production patterns.

This goal is pursued through 11 targets (Arcuri & Partiti, 2021; Shulla & Filho, 2023). These targets are as follows (United Nations, 2015):

- 12.1 Implement the 10-Year Framework of Programmes on Sustainable Consumption and Production Patterns, all countries taking action, with developed countries taking the lead, taking into account the development and capabilities of developing countries.
- 12.2 By 2030, achieve the sustainable management and efficient use of natural resources.
- 12.3 By 2030, halve per capita global food waste at the retail and consumer levels and reduce food losses along production and supply chains, including post-harvest losses.
- 12.4 By 2020, achieve the environmentally sound management of chemicals and all wastes throughout their life cycle, in accordance with agreed international frameworks, and significantly reduce their release to air, water and soil in order to minimize their adverse impacts on human health and the environment.
- 12.5 By 2030, substantially reduce waste generation through prevention, reduction, recycling, and reuse.
- 12.6 Encourage companies, especially large and transnational companies, to adopt sustainable practices and to integrate sustainability information into their reporting cycle.
- 12.7 Promote public procurement practices that are sustainable, in accordance with national policies and priorities.
- 12.8 By 2030, ensure that people everywhere have the relevant information and awareness for sustainable development and lifestyles in harmony with nature.
- 12.a Support developing countries to strengthen their scientific and technological capacity to move towards more sustainable patterns of consumption and production.
- 12.b Develop and implement tools to monitor sustainable development impacts for sustainable tourism that creates jobs and promotes local culture and products.
- 12.c Rationalize inefficient fossil-fuel subsidies that encourage wasteful consumption by removing market distortions, in accordance with national circumstances, including by restructuring taxation and phasing out those harmful subsidies, where they exist, to reflect their environmental impacts, taking fully into account the specific needs and conditions of developing countries and minimizing the possible adverse impacts on their development in a manner that protects the poor and the affected communities.

In the case of Goal 12, sustainable consumption and production can be considered separately or together. During the production process, resources, energy, parts, packaging, and services are utilized (consumed). However, mass consumption is responsible for the production of goods (Glavič, 2021). The issues regarding sustainable production were discussed in Chapter 5 of this book. Unfortunately, progress in implementing Goal 12 of sustainable consumption and production has been limited and slow (Guevara & Julián, 2019).

11.3 Sustainable and unsustainable consumption of products

Two approaches to sustainable consumption can be distinguished in the literature: strong and weak (Neale, 2015). Weak sustainable consumption is rooted in market-based approaches and relies on technological optimism and focuses on promoting more efficient production methods and products primarily through technological progress and conscious consumer choice (Bengtsson et al., 2018). However, this approach has limited potential when it comes to addressing contemporary sustainability challenges, such as the lack of attention paid to such issues as justice, the inability to deal with rebound effects, and the neglect of general constraints (Lorek & Fuchs, 2013). On the other hand, strong sustainable consumption underscores the crucial role played by social innovation as a starting point for achieving this objective (Lorek & Spangenberg, 2014). It highlights the need for changes in the volume and patterns of consumption (Lorek & Fuchs, 2013).

One of the trends shaping the sustainable consumption of products is the reduced quantity of purchased and used products, this phenomenon referred to as deconsumption.

Deconsumption is an alternative approach that involves limiting the consumption of products or the utilization of goods. For the most part, this entails people making modifications to their lifestyle and the ways in which they satisfy their own needs (Patrzałek, 2019).

The approach that involves making consumption decisions based more on sustainable ecological development is called green consumption (Goodwin et al., 2020). Green consumption involves choosing products with minimal environmental impact, moderate usage, increased awareness of waste generation, collaboration with environmentally friendly companies, frequent recycling of products, and opting for clean or renewable energy sources (Haider et al., 2022; Lu et al., 2013; Semprebon et al., 2019). Two basic types of green consumption can be distinguished (Goodwin et al., 2020):

1 Shallow consumption, where consumers aim to buy "environmentally friendly" products but may not necessarily change their overall consumption levels. However, according to Cheng et al. (2020) and Muldoon (2006), even in this approach, the level of green consumption of products and services may increase.
2 Deep consumption, where consumers strive not only to purchase environmentally friendly products but also, and more importantly, aim to reduce their overall consumption levels.

Another trend favouring sustainable consumption is consumer ethnocentrism. This is the tendency of consumers to prefer purchasing locally produced products over foreign alternatives (Jiménez & San Martín, 2010; Ma et al., 2019). Consumer ethnocentrism is typically considered at the national level. However, it can also be examined at the local level (Bryła, 2019; Siamagka & Balabanis, 2015). According to Fernández-Ferrín and Bande-Vilela (2015), consumer ethnocentrism can be considered from three perspectives, namely:

• Transnational (e.g., consumers who identify with the European Union),
• National, and
• Sub-national (regional and local).

In their study, Lee et al. (2015) confirmed the thesis that environmental attitudes and sustainable consumption behaviour is enhanced by consumption values centred on a strong identity with place. Residents who enjoy living in their community are more positive about sustainable consumption.

One concept that is closely related to sustainable consumption is the circular economy (CE). In the transition to a CE, extending the life cycle of products used by consumers plays an important role. This is accomplished by preventing wastage of resources (Kurzak-Mabrouk, 2019) and adopting the R practices. 3R practice (Reduce, Reuse, Recycle) is based on the implementation of the postulate of conscious consumption. Its aim is to limit the purchases made by consumers to products they regard as necessary. Moreover, consumers should use such products repeatedly and, as a result, process them properly (Patwa et al., 2021). In recent years, the R methodology has been further developed. This led to the addition of many R components to the mentioned 3R (Skärin et al., 2022). The authors (Potting et al., 2017) developed the methodology for 10R (Refuse, Rethink, Reduce, Reuse, Repair, Refurbish, Remanufacture, Repurpose, Recycle, and Recover). It is worth pointing out that conscious consumption promotes greater consumer empowerment with an enhanced ability to recognize consumption patterns and ways of minimizing the negative effects of consumption via sustainable development (da Silva Nascimento et al., 2018; Silva et al., 2012).

The opposite approach to sustainable consumption is unsustainable consumption. Unsustainable consumption has two sides; on one hand, it concerns excessive consumption of products and services, which is typical of rich countries. On the other hand, insufficient consumption, that is, below the level that enables members of poor societies to meet their basic needs, predominates in poorer societies.

One example of insufficient consumption is inadequate consumption of fruits and vegetables in individuals aged 15 and older in 28 low- and middle-income countries (Siegel, 2019).

One of the main causes of unsustainable consumption is believed to be consumerism. However, it is not the sole cause (Holt, 2012). One major problem that has not been sufficiently resolved is increased overconsumption resulting from falling product prices. The implication is that global capital mobility and an excessive global supply of labour have allowed companies to lower wages and avoid incurring environmental costs. This allows consumers to buy more of these cheap goods

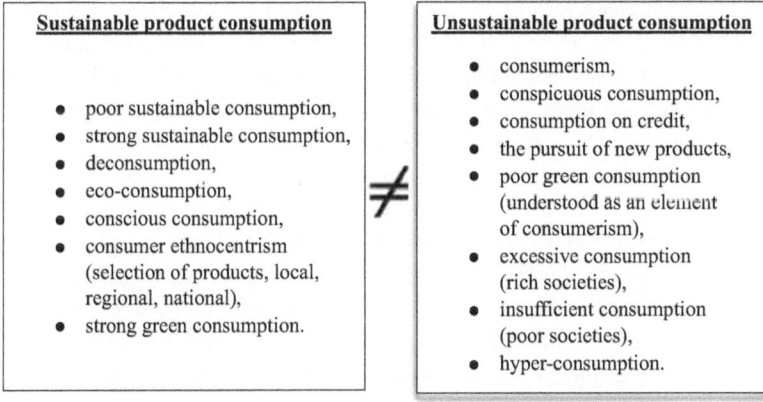

Sustainable product consumption	Unsustainable product consumption
• poor sustainable consumption, • strong sustainable consumption, • deconsumption, • eco-consumption, • conscious consumption, • consumer ethnocentrism (selection of products, local, regional, national), • strong green consumption.	• consumerism, • conspicuous consumption, • consumption on credit, • the pursuit of new products, • poor green consumption (understood as an element of consumerism), • excessive consumption (rich societies), • insufficient consumption (poor societies), • hyper-consumption.

FIGURE 11.1   Features and trends of sustainable and unsustainable product consumption.
Source: Authors' own table.

(Schor, 2005). A very dangerous trend in consumerism is hyper-consumption. This consumption model has a number of negative consequences both for consumers at the local level and for the environment and society at the global level.

Among the negative effects of hyper-consumption, researchers (Angelova et al., 2021) point out the following: a high volume of leftovers and food waste, obesity, credit debt, a lack of savings, the purchase of new electrical appliances due to a reluctance to repair existing ones, and hyper mobility. Hyper-consumption, which poses the greatest threat to the environment and humanity, is best addressed by two different sustainability strategies. These strategies include (Dhandra, 2019):

- encouraging environmentally friendly or pro-ecological consumption (Stern, 2000),
- encouraging restraint in consumption or reducing consumption by limiting the exploitation of natural resources and improving prosperity (Sheth et al., 2011).

The characteristics and trends of sustainable and unsustainable product consumption are shown in Figure 11.1.

11.4   Challenges to implementing sustainable product consumption

In today's society, consumption is becoming an increasingly significant aspect of life. When consumption is unsustainable, it leads to adverse consequences for the environment and society. Research indicates that the current ecological and environmental challenges facing society have been exacerbated by population growth and excessive consumption (Chen & Hung, 2016).

To reduce the negative effects of unsustainable consumption, a more conscious approach to shopping needs to be promoted. Therefore, educating consumers about ecological alternatives, recycling, and waste minimization is vital. Companies should also shift to more sustainable production and sourcing practices, cutting back on the use of natural resources, reducing emissions, and providing employees with decent working conditions and suitably high wages.

In addition to the positive aspects, there are a number of challenges and barriers to implementing sustainable consumption. The limitations include the fact that manufacturing of sustainable products may entail higher costs, which may affect the final price for consumers. This may make sustainable products less competitive. Other barriers include a lack of awareness of or psychological resistance to changing consumer habits. These barriers may make it difficult for consumers to adopt a sustainable consumption model.

An article by Sundaraja et al. (2021) presents research conducted on a group of 781 Australian consumers regarding the purchase of products containing palm oil obtained from sustainable crops. It turned out that almost half of them had never bought this type of product. The researchers identified that the following major barriers to buying these sustainable products: a lack of knowledge about the problems associated with palm oil production, uncertainty about product availability, and poor attitudes towards green consumption.

As research by Steinbiß et al. (2022) shows, environmental motivations are usually secondary to self-interest. Therefore, they argue

> ...it seems necessary to focus empirical results on the most important aspects of changing consumer purchasing behaviour towards sustainability in a specific industry. Individual self-interest is the strongest driver of sustainable consumer purchasing behaviour. The idea of doing something good for your body, e.g. by eating foods rich in vitamins, introduces sustainable products into purchasing behaviour....

In turn, according to other research, the main factors shaping ecological purchasing intentions are cultural values and ecological advertising, while there is no strong correlation between knowledge about the environment and a willingness to make ecological purchases. At the same time, it was determined that the motivational factors for green purchasing intentions are greater among people with a higher education, especially women (Chekima et al., 2016).

11.5 Consumer expectations of products in the context of sustainable consumption

From the consumer's point of view, the most important thing is that a product of appropriate quality meets his or her expectations and preferences and is on offer at an appropriate price that is competitive with similar products. A number of other aspects of a product are also important for the consumer, such as the expected length of use, the length of the warranty, the ease and costs of its servicing, and the costs of exploitation. At the same time, due to increased ecological and social awareness, consumers increasingly expect products to be ecologically and socially sustainable, so that they can practice sustainable consumption.

From a manufacturer's point of view, it is important to offer products of appropriate quality that meet customer expectations, are produced at the lowest possible cost, and are sold at the highest price acceptable to consumers. In light of public opinion (associations, organizations, movements, and non-governmental sector) and the requirements of authorities at various levels (national, European, and international), products on offer are expected to be, among other things, innovative, ecological, and sustainable.

These entities are also increasingly demanding that enterprises manufacture products within the CE. The most important consumer expectations of stakeholders regarding sustainable products are presented in Figure 11.2.

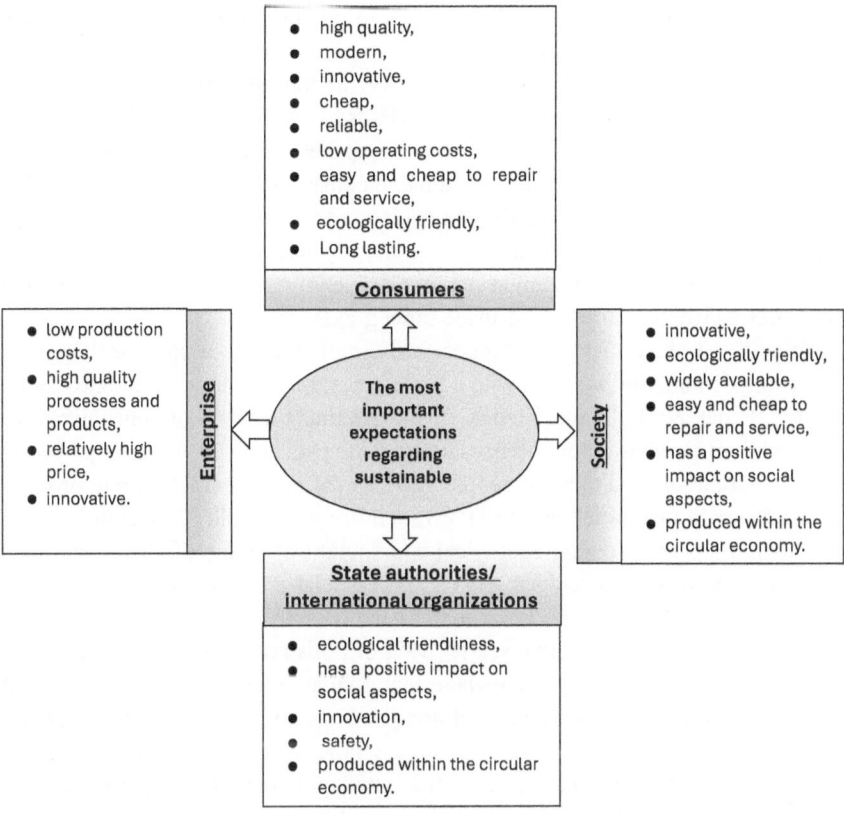

FIGURE 11.2   The most important expectations of stakeholders regarding sustainable products.

Source: Author's own diagram.

It is practically impossible to meet all the possible expectations of the various stakeholders. Therefore, it is necessary to balance features and expectations in relation to the product. The concept of a sustainable product responds well to this challenge.

Sustainable products not only have a positive impact on the immediate environment and people directly involved in the product life cycle (e.g., suppliers, employees, customers) but also take into account the indirect impact of the product life cycle on society and ecosystems (Dyllick & Rost, 2017). Product sustainability may encompass a variety of legal, cultural, ethical, climatic, and political issues. However, product sustainability is usually considered in terms of its economic, social, and ecological aspects (Rosen & Kishawy, 2012).

The need for sustainable development was first noticed by ecological and social organizations, movements, and associations. Society's growing awareness of the need for sustainable development is being reflected in measures taken by international and national authorities (in the form of acts, directives, resolutions, regulations, recommendations, standards, etc.), encouraging companies to design and produce sustainable products and customers to practice sustainable consumption. At the same time, the growing awareness of the need for sustainable consumption, thanks to the influence exerted by both society and the authorities, means that consumers increasingly expect sustainable products.

The author of the article (Hale, 2018) believes that consumers' ecological awareness is the main factor increasing sustainable consumption and policies should be strengthened in this area. A similar claim is made by Stern (2000), according to whom pro-ecological behaviour (e.g., purchasing decisions) encourages companies to produce socially and environmentally friendly goods and at the same time motivates consumers to purchase these same goods.

Other researchers highlight, supported by marketing research, that there is an increasing demand for products with a socially responsible or pro-ecological brand image (Niedek & Hoffmann-Niedek, 2014). As many as 79% of consumers have changed their purchasing preferences based on the social or environmental impact of their purchases, and 57% of people aged 18–24 have decided to buy products from a lesser-known brand because they are more sustainable (Capgemini, 2020).

This reflects the growing ecological and social awareness of customers, who, influenced by society, authorities, and various associations and movements, increasingly expect products to be environmentally friendly and socially responsible (Tu et al., 2018). According to research, 19% of Polish respondents and an average of 27% of respondents worldwide claim that they are ready to pay more for products manufactured in compliance with the principles of sustainable development. At the same time, however, as many as 90% of Poles claim they are discouraged by the high prices of such products (EY Polska, 2021). In turn, 75% of Polish fashion industry consumers claim that sustainable development is an important factor in their everyday lives, including when making purchasing decisions. However, study showed that when the price of sustainability increased by 20%, customer demand

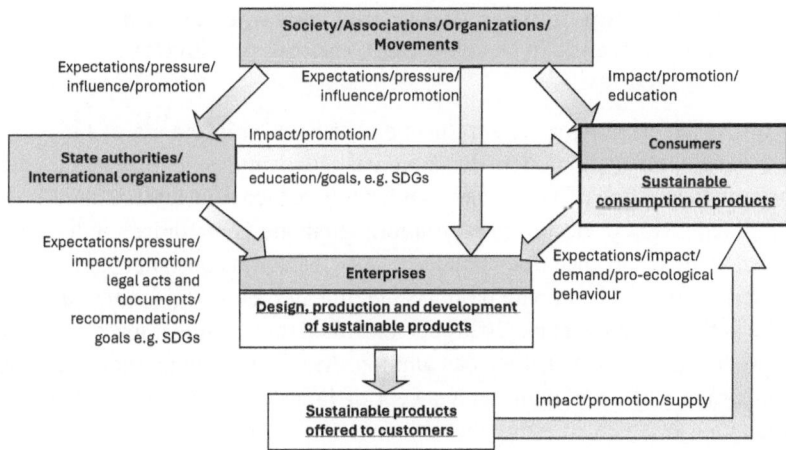

FIGURE 11.3   Stakeholder influence on the consumption, design, and production of sustainable products.

Source: Author's own diagram.

dropped by 50% for sustainable shirts and by 62% for sustainable shoes. And when prices rose by 50%, customers' willingness to buy both types of sustainable product dropped to almost zero (Vogue Polska & BCG, 2021).

It is also worth noting differences in the willingness of different generations to pay an additional price for sustainable products. According to Tidswell (2023), research conducted among American consumers found that 54% of Generation Z consumers, 50% of Millennials, and only 34% of Baby Boomers consumers were willing to pay 10% more for sustainable products. As can be seen, the concept of sustainable consumption is most closely aligned with young consumers from Generation Z.

All these expectations, regulations and influences mean that companies must increasingly take into account ecological and social factors when designing, producing, developing, and managing products. It seems that the pressure (influence) exerted on entrepreneurs to produce sustainable products will continue to grow. Moreover, thanks to the current availability of sustainable products, combined with their promotion and advertising, an increasing number of consumers will expect this type of product in the future (Figure 11.3).

The study's authors emphasize the fact that social media marketing and the use of social media play a key role in shaping consumers' purchasing intentions when it comes green products and services, which in turn lead to increased sustainable consumption (Nekmahmud et al., 2022). At the same time, when ecological and social factors are taken even more into account as part of the sustainability process consumers will be more willing to pay a premium for sustainable product options.

11.6 Relationship between sustainable consumption and modern trends in development, design, production, and business

One of the factors supporting sustainable consumption in the era of Industry 4.0 is the development of modern technologies. Innovations in renewable energy, energy efficiency, and recycling can significantly reduce the negative impact on the environment. Moreover, advances in information and communication technology have led to the emergence of e-commerce, which enables consumers to make more informed choices and compare products in terms of their sustainability. In the article, the authors (Saniuk et al., 2020) demonstrated the positive impact of personalized production on sustainable consumption. At the same time, they emphasized the importance of developing the concept of Industry 4.0 to support sustainable consumption-oriented consumer behaviour.

In turn, other researchers (Hengboriboon et al., 2022) examined the impact on consumer decision-making of product image and the corporate reputation of organizations implementing corporate social responsibility (CSR) programmes. Their research showed that product image and company reputation were key factors shaping the perspective and purchasing intentions of Thai consumers regarding ecological food.

Striving for sustainable product consumption is essential for achieving the goals of the CE and sustainable development. By promoting sustainable practices in the design, production, and consumption of products, we can aim to create a more sustainable and equitable society that cares for future generations while also protecting the environment. On the other hand, modern trends in product development and design, sustainable manufacturing and business, the CE, Industry 4.0, and CSR support the implementation and achievement of sustainable consumption goals, as depicted in Figure 11.4.

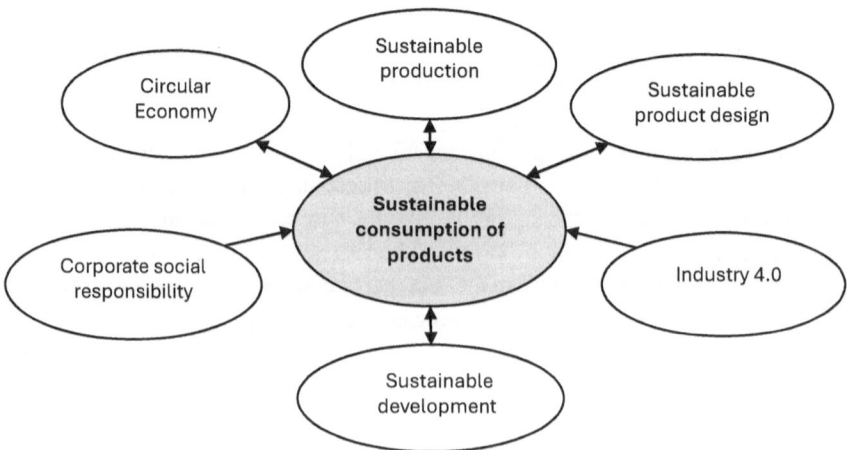

FIGURE 11.4   Connections between sustainable consumption and modern trends in development, design, production, and business.

Source: Author's own diagram.

11.7   Conclusions

This chapter discusses sustainable product consumption as a component of sustainable development. It then presents the characteristics and trends of sustainable and unsustainable product consumption. Furthermore, it describes consumer expectations of products in the context of sustainable consumption. It also outlines the challenges involved in implementing sustainable product consumption. This chapter illustrates the impact of stakeholders on the design, production, and consumption of sustainable products. It also discusses the connections between sustainable consumption and modern trends in development, design, production, and business. To summarize this chapter, the following conclusions (in terms of implementation) can be drawn:

- In the face of accelerating climate change, increasing environmental pollution, and excessive use of natural resources, the issue of sustainable product consumption is becoming increasingly important. Decisions made by various stakeholders, from producers to consumers to politicians, have a huge impact on the direction in which contemporary consumption is developing and, consequently, on the state of the environment and society as a whole.
- The sustainable products offered by businesses, coupled with their promotion and advertising, have helped increase the number of consumers expecting such products.
- Sustainable consumption should be promoted, supported, and implemented at every level, whether globally, via the United Nations, internationally, for example through the European Union, nationally, regionally, or locally. Every consumer should strive to increase their sustainable consumption through the purchase, consumption, and repair of products that ensures their multiple usage, and then recycling in an appropriate fashion.
- Effective methods for shifting consumer attitudes away from consumerism and towards sustainable product consumption include the following: raising ecological awareness, advertising and promotion, especially in social media, and developing modern technologies as part of the concept of Industry 4.0.
- Modern trends in sustainable development, design, manufacturing, and sustainable business, along with the concepts of Industry 4.0, the CE, and CSR, favour the implementation and achievement of the goals of sustainable consumption.

References

Angelova, M., Dimitrova, T., & Pastarmadzhieva, D. (2021). The effects of globalization: Hyper consumption and environmental consumer behavior during the Covid-19 pandemic. *International Journal of Economics and Business Administration*, *IX*(Issue 4). https://doi.org/10.35808/ijeba/733

Arcuri, A., & Partiti, E. (2021). *TILEC Discussion Paper. SDG 12: Ensure Sustainable Consumption and Production Patterns*. https://ssrn.com/abstract=3814765

Becker, K. (2021). Raport „Vogue Polska" i BCG: Polscy konsumenci a zrównoważony rozwój mody. https://www.vogue.pl/a/raport-vogue-polska-i-bcg-polscy-konsumenci-a-zrownowazony-rozwoj-mody (Accessed on 7 October 2024).

Bengtsson, M., Alfredsson, E., Cohen, M., Lorek, S., & Schroeder, P. (2018). Transforming systems of consumption and production for achieving the sustainable development goals: moving beyond efficiency. *Sustainability Science*, *13*(6). https://doi.org/10.1007/s11625-018-0582-1

Brundtland, G. (1987). UN Brundtland Commission Report. Our Common Future.

Bryła, P. (2019). Regional ethnocentrism on the food market as a pattern of sustainable consumption. *Sustainability (Switzerland)*, *11*(22). https://doi.org/10.3390/su11226408

Capgemini. (2020, October 16). *Jak zrównoważony rozwój wpływa na decyzje zakupowe klientów?* https://Capgeminipolska.Prowly.Com/107835-Jak-Zrownowazony-Rozwoj-Wplywa-Na-Decyzje-Zakupowe-Klientow (Accessed on 14 March 2024).

Chekima, B., Chekima, S., Syed Khalid Wafa, S. A. W., Igaua, O. A., & Sondoh, S. L. (2016). Sustainable consumption: The effects of knowledge, cultural values, environmental advertising, and demographics. *International Journal of Sustainable Development and World Ecology*, *23*(2). https://doi.org/10.1080/13504509.2015.1114043

Chen, S. C., & Hung, C. W. (2016). Elucidating the factors influencing the acceptance of green products: An extension of theory of planned behavior. *Technological Forecasting and Social Change*, *112*. https://doi.org/10.1016/j.techfore.2016.08.022

Cheng, Z. H., Chang, C. T., & Lee, Y. K. (2020). Linking hedonic and utilitarian shopping values to consumer skepticism and green consumption: The roles of environmental involvement and locus of control. *Review of Managerial Science*, *14*(1). https://doi.org/10.1007/s11846-018-0286-z

da Silva Nascimento, L., Ferraz Soares de Lima, L., & Vicente Sales Melo, F. (2018). Collaborative consumption: A quantitative research in light of the conscious consumption of car sharing users. *Journal of Marketing Management*, *6*(1), 41–54.

Dhandra, T. K. (2019). Achieving triple dividend through mindfulness: More sustainable consumption, less unsustainable consumption and more life satisfaction. *Ecological Economics*, *161*. https://doi.org/10.1016/j.ecolecon.2019.03.021

Dyllick, T., & Rost, Z. (2017). Towards true product sustainability. *Journal of Cleaner Production*, *162*. https://doi.org/10.1016/j.jclepro.2017.05.189

EY Polska. (2021, December 7). *Czwarta polska edycja badania EY Future Consumer Index. Pandemia zwiększyła zainteresowanie konsumentów zrównoważonym rozwojem.* https://Www.Ey.Com/Pl_pl/News/2021/12/Ey-Future-Consumer-Index-2021-Zrownowazony-Rozwoj (Accessed on 14 March 2024).

Fernández-Ferrín, P., & Bande-Vilela, B. (2015). Attitudes and reactions of Galician (Spanish) consumers towards the purchase of products from other regions. *Global Business and Economics Review*, *17*(2). https://doi.org/10.1504/GBER.2015.068563

Glavič, P. (2021). Evolution and current challenges of sustainable consumption and production. *Sustainability (Switzerland)*, *13*(16). https://doi.org/10.3390/su13169379

Goodwin, N., Harris, J., Nelson, J., Roach, B., & Torras, M. (2020). Consumption and the consumer society. *Microeconomics in Context*. https://doi.org/10.4324/9781315702414-19

Guevara, S., & Julián, I. P. (2019). Sustainable consumption and production: A crucial goal for sustainable development—Reflections on the Spanish SDG implementation report. *Journal of Sustainability Research*, *1*(2). https://doi.org/10.20900/jsr20190019

Haider, M., Shannon, R., & Moschis, G. P. (2022). Sustainable consumption research and the role of marketing: A review of the literature (1976–2021). *Sustainability (Switzerland)*, *14*(7). https://doi.org/10.3390/su14073999

Hale, L. A. (2018). At home with sustainability: From green default rules to sustainable consumption. *Sustainability (Switzerland)*, *10*(1). https://doi.org/10.3390/su10010249

Hengboriboon, L., Naruetharadol, P., Ketkeaw, C., & Gebsombut, N. (2022). The impact of product image, CSR and green marketing in organic food purchase intention: Mediation roles of corporate reputation. *Cogent Business and Management, 9*(1). https://doi.org/ 10.1080/23311975.2022.2140744

Holt, D. B. (2012). Constructing sustainable consumption: From ethical values to the cultural transformation of unsustainable markets. *Annals of the American Academy of Political and Social Science, 644*(1). https://doi.org/10.1177/0002716212453260

Iwasiński, Ł. (2014). Co to znaczy konsumować? Próba definicji pojęcia konsumpcji. *Konsumpcja i Rozwój, 4*(9), 14–23.

Jiménez, N. H., & San Martín, S. (2010). The role of country-of-origin, ethnocentrism and animosity in promoting consumer trust. The moderating role of familiarity. *International Business Review, 19*(1). https://doi.org/10.1016/j.ibusrev.2009.10.001

Kuhlman, T., & Farrington, J. (2010). What is sustainability? *Sustainability, 2*(11). https://doi.org/10.3390/su2113436

Kurzak-Mabrouk, A. (2019). Niezrównoważona konsumpcja i sposoby jej równoważenia. *Forum Socjologiczne, 9*. https://doi.org/10.19195/2083-7763.9.3

Lee, C. K. C., Levy, D. S., & Yap, C. S. F. (2015). How does the theory of consumption values contribute to place identity and sustainable consumption? *International Journal of Consumer Studies, 39*(6). https://doi.org/10.1111/ijcs.12231

Lorek, S., & Fuchs, D. (2013). Strong sustainable consumption governance – Precondition for a degrowth path? *Journal of Cleaner Production, 38*. https://doi.org/10.1016/j. jclepro.2011.08.008

Lorek, S., & Spangenberg, J. H. (2014). Sustainable consumption within a sustainable economy – beyond green growth and green economies. *Journal of Cleaner Production, 63*, 33–44. https://doi.org/10.1016/J.JCLEPRO.2013.08.045

Lu, L., Bock, D., & Joseph, M. (2013). Green marketing: What the Millennials buy. *Journal of Business Strategy, 34*(6). https://doi.org/10.1108/JBS-05-2013-0036

Ma, Q., Abdeljelil, H. M., & Hu, L. (2019). The influence of the consumer ethnocentrism and cultural familiarity on brand preference: Evidence of event-related potential (ERP). *Frontiers in Human Neuroscience, 13*. https://doi.org/10.3389/fnhum.2019.00220

Mont, O., & Plepys, A. (2008). Sustainable consumption progress: Should we be proud or alarmed? *Journal of Cleaner Production, 16*(4). https://doi.org/10.1016/j.jclepro.2007.01.009

Muldoon, A. (2006). Where the green is: Examining the paradox of environmetally conscious consumption. *Electronic Green Journal, 23*. https://doi.org/10.5070/g312310643

Neale, A. (2015). Zrównoważona konsumpcja. Źródła koncepcji i jej zastosowania. *Prace Geograficzne, 141*, 141–158.

Nekmahmud, M., Naz, F., Ramkissoon, H., & Fekete-Farkas, M. (2022). Transforming consumers' intention to purchase green products: Role of social media. *Technological Forecasting and Social Change, 185*. https://doi.org/10.1016/j.techfore.2022.122067

Niedek, M., & Hoffmann-Niedek, A. (2014). Produkcja ekologiczna zrównoważona w świetle odpowiedzialności biznesu. *Optimum. Studia Ekonomiczne, 4*(70), 46–60.

Ofstad, S., Westly, L., & Bratelli, T. (1994). Symposium: sustainable consumption. Ministry of the Environment, Oslo

Patrzałek, W. (2019). Between consumerism and deconsumption – in search of a new model of society. *Prace Naukowe Uniwersytetu Ekonomicznego We Wrocławiu, 63*(10). https://doi.org/10.15611/pn.2019.10.16

Patwa, N., Sivarajah, U., Seetharaman, A., Sarkar, S., Maiti, K., & Hingorani, K. (2021). Towards a circular economy: An emerging economies context. *Journal of Business Research, 122*. https://doi.org/10.1016/j.jbusres.2020.05.015

Potting, J., Hekkert, M., Worrell, E., & Hanemaaijer, A. (2017). Circular economy: Measuring innovation in the product chain. Policy Report. *PBL Netherlands Environmental Assessment Agency, 2544*, 42 p.

Ramkissoon, H., Graham Smith, L. D., & Weiler, B. (2013). Testing the dimensionality of place attachment and its relationships with place satisfaction and pro-environmental behaviours: A structural equation modelling approach. *Tourism Management, 36*. https://doi.org/10.1016/j.tourman.2012.09.003

Rosen, M. A., & Kishawy, H. A. (2012). Sustainable manufacturing and design: Concepts, practices and needs. *Sustainability, 4*(2). https://doi.org/10.3390/su4020154

Saniuk, S., Grabowska, S., & Gajdzik, B. Z. (2020). Personalization of products in the industry 4.0 concept and its impact on achieving a higher level of sustainable consumption. *Energies, 13*(22). https://doi.org/10.3390/en13225895

Schor, J. B. (2005). Prices and quantities: Unsustainable consumption and the global economy. *Ecological Economics, 55*(3). https://doi.org/10.1016/j.ecolecon.2005.07.030

Semprebon, E., Mantovani, D., Demczuk, R., Souto Maior, C., & Vilasanti, V. (2019). Green consumption: A network analysis in marketing. *Marketing Intelligence and Planning, 37*(1). https://doi.org/10.1108/MIP-12-2017-0352

Sheth, J. N., Sethia, N. K., & Srinivas, S. (2011). Mindful consumption: A customer-centric approach to sustainability. *Journal of the Academy of Marketing Science, 39*(1). https://doi.org/10.1007/s11747-010-0216-3

Shulla, K., & Filho, W. L. (2023). Achieving the UN Agenda 2030: Overall actions for the successful implementation of the Sustainable Development Goals before and after the 2030 deadline. European Union Parliament.

Siamagka, N. T., & Balabanis, G. (2015). Revisiting consumer ethnocentrism: Review, reconceptualization, and empirical testing. *Journal of International Marketing, 23*(3). https://doi.org/10.1509/jim.14.0085

Siegel, K. R. (2019). Insufficient consumption of fruits and vegetables among individuals 15 years and older in 28 low- and middle-income countries: What can be done? *Journal of Nutrition, 149*(7). https://doi.org/10.1093/jn/nxz123

Silva, M. das G. e, Araújo, N. M. S., & Santos, J. S. (2012). "Consumo consciente": o ecocapitalismo como ideologia. *Revista Katálysis, 15*(1). https://doi.org/10.1590/s1414-49802012000100010

Skärin, F., Rösiö, C., & Andersen, A. L. (2022). An explorative study of circularity practices in Swedish manufacturing companies. *Sustainability (Switzerland), 14*(12). https://doi.org/10.3390/su14127246

Steinbiß, K., Fröhlich, E., & Sander, J. (2022). Managing sustainable consumption: Shaping the customer journey with focus on sustainability in the food industry. *Edukacja Ekonomistów i Menedżerów, 62*(4). https://doi.org/10.33119/eeim.2021.62.4

Stern, P. C. (2000). Toward a coherent theory of environmentally significant behavior. *Journal of Social Issues, 56*(3). https://doi.org/10.1111/0022-4537.00175

Strezov, V., Evans, A., & Evans, T. J. (2017). Assessment of the economic, social and environmental dimensions of the indicators for sustainable development. *Sustainable Development, 25*(3). https://doi.org/10.1002/sd.1649

Sundaraja, C. S., Hine, D. W., & Lykins, A. D. (2021). Palm oil: Understanding barriers to sustainable consumption. *PLoS One, 16*(8 August). https://doi.org/10.1371/journal.pone.0254897

Tidswell, E. (2023). Does Gen Z care about sustainability? Stats & facts in 2023. Goodmakertales.Com (Accessed on 14 March 2024).

Tu, J. C., Zhang, X. Y., & Huang, S. Y. (2018). Key factors of sustainability for smartphones based on Taiwanese consumers' perceived values. *Sustainability (Switzerland), 10*(12). https://doi.org/10.3390/su10124446

United Nations. (2015). Transforming our world: The 2030 Agenda for Sustainable Development. United Nations Sustainable knowledge platform. Sustainable Development Goals.

Vargas-Merino, J. A., Rios-Lama, C. A., & Panez-Bendezú, M. H. (2023). Sustainable consumption: Conceptualization and characterization of the complexity of "being" a sustainable consumer—A systematic review of the scientific literature. *Sustainability (Switzerland)*, *15*(10). https://doi.org/10.3390/su15108401

Voulvoulis, N., Giakoumis, T., Hunt, C., Kioupi, V., Petrou, N., Souliotis, I., Vaghela, C., & binti Wan Rosely, W. I. H. (2022). Systems thinking as a paradigm shift for sustainability transformation. *Global Environmental Change*, *75*. https://doi.org/10.1016/j.gloenvcha.2022.102544

Wang, C., Ghadimi, P., Lim, M. K., & Tseng, M. L. (2019). A literature review of sustainable consumption and production: A comparative analysis in developed and developing economies. *Journal of Cleaner Production*, *206*. https://doi.org/10.1016/j.jclepro.2018.09.172

Zalega, T. (2020). Sustainable consumption in consumer behaviour of young Polish singles. *Acta Scientiarum Polonorum. Oeconomia*, *19*(1). https://doi.org/10.22630/aspe.2020.19.1.10

12

SUSTAINABLE PRODUCT INNOVATION

Biopolymers as a case study

Tomasz Witko, Ignazio Blanco and Karolina E. Mazur

12.1   Introduction

As a group of materials with a relatively short history, plastics have revolutionized global industry. Materials have shaped humanity since the dawn of time, through the Bronze and Iron ages to Ancient Rome, the Industrial Revolutions, and now in the Era of Technology, often referred to in materials science as the "Era of Composites". The latter revolution began with the introduction of a completely synthetic material developed by Leo Baekeland in 1907, which came to be known as Bakelite in his honor. Another synthetic polymer material was polyethylene (PE), which was first synthesized in 1898 by Hans von Pechmann and whose processing properties were improved by subsequent scientists thanks to which mass production of this material was possible by the second half of the twentieth century (Chakrabarty et al., 2015; Lüftl et al., 2016). A milestone in plastics production came with the synthesis of polyethylene terephthalate (PET) achieved by James Tennant Dickson and John Rex Whinfield, which revolutionized the packaging industry and replaced materials such as glass, cardboard, aluminum, and cellulose itself (Montava-Jordà et al., 2019).

Plastics have gained in popularity partly due to its ability to be mass produced but also because of its expected properties, primarily its weight compared to the products dimensions themselves. Scientists' efforts to improve plastics have led to advances in almost every industry. There is a market for plastics in every trade sector, with the packaging industry boasting the highest share of total plastics consumption, followed by construction, textiles, automotive, transport, and consumer goods. Global plastic production was estimated at 420 million metric tons in 2022 (PlasticEurope: Plastics – The Facts 2022, 2024).

Plastic materials possess a wide range of properties. They are designed to meet the needs of each single application in the most efficient manner possible.

DOI: 10.4324/9781032710693-13

Currently, most plastic materials are derived from fossil feedstocks such as gas, oil, or coal. However, it should be pointed out that only 8% of all the oil and gas used in Europe is utilized in the production of plastic materials (Prieto, 2007; "Winter Is Coming: Plastic Has To Go: A Case for Decreasing Plastic Production to Reduce the European Union's Dependence on Fossil Fuels and Russia", 2022).

The increasing demand for affordable and resilient plastics is contributing to a rapidly escalating global waste crisis (Laird, 2022). Numerous technologies are currently being developed to tackle the issue of plastic waste, with a focus on either energy recovery or mechanical recycling. Energy recovery involves closing the multimillion-year life cycle of plastics by thermally decomposing them into CO_2 or liquid fuel, which can later be converted back into CO_2 through combustion. However, one drawback of mechanical recycling is that it tends to compromise the mechanical properties of the material, making the process unsustainable in the long run (Nitkiewicz et al., 2020). The table below (Figure 12.1) shows the carbon lifecycle, with the rates of exchanges between individual stages color marked. It can be clearly seen that the application of bio-polymers, such as polyhydroxyalkanoates (PHAs) and polylactide (PLA), eliminates the lengthy composting process and prevents the accumulation of materials in landfills (Siracusa et al., 2020).

Mineralization of organic matter into fossil fuels is a very slow process. Crude oil is quickly recovered and can be easily converted into fuels or traditional plastics. Plastic waste is produced at a very fast rate and its mineralization subsequently takes hundreds of years. Plastic waste can serve as a microbial fermentation substrate, which is converted into PHA, thus enabling capture of the CO_2. Biopolymers undergo natural decomposition and can be composted, which results in much slower and limited CO_2 release into the atmosphere.

Petrochemical plastics exhibit inherent resistance to biodegradation, and the widespread use of conventional polymers derived from petrochemical sources has yet to demonstrate extensive biodegradability. The staggering scale of plastics manufacturing and consumption, combined with inadequate disposal practices has given rise to a pressing global waste management crisis. Recent data underscores the severity of this problem, revealing that in 2022, out of the more than 300 million tons of waste generated in the United States, nearly one-fifth comprised plastics. Alarmingly, the proportion of plastics in landfills continues to escalate, signaling a growing environmental challenge (Hottle et al., 2017).

Less than 10% of the plastic waste produced in the United States in 2022 was recovered for recycling purposes, leaving a substantial 55.92 million tons destined for landfill or incineration. These statistics underscore the urgency of adopting more sustainable practices in plastic usage, disposal, and recovery efforts. The data highlights the paramount need for innovative solutions and a paradigm shift in waste management strategies to mitigate the escalating environmental impact of petrochemical plastics on a global scale. Although, the European situation is improving, in many countries, landfills are still the first or second option of treatment for plastic post-consumer waste. In countries such as Switzerland, Austria,

A

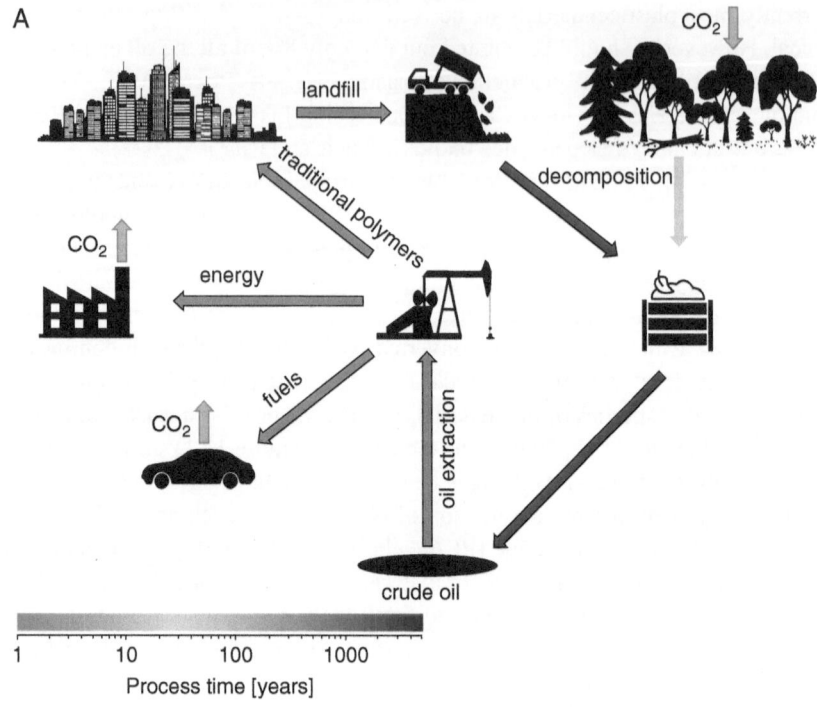

B

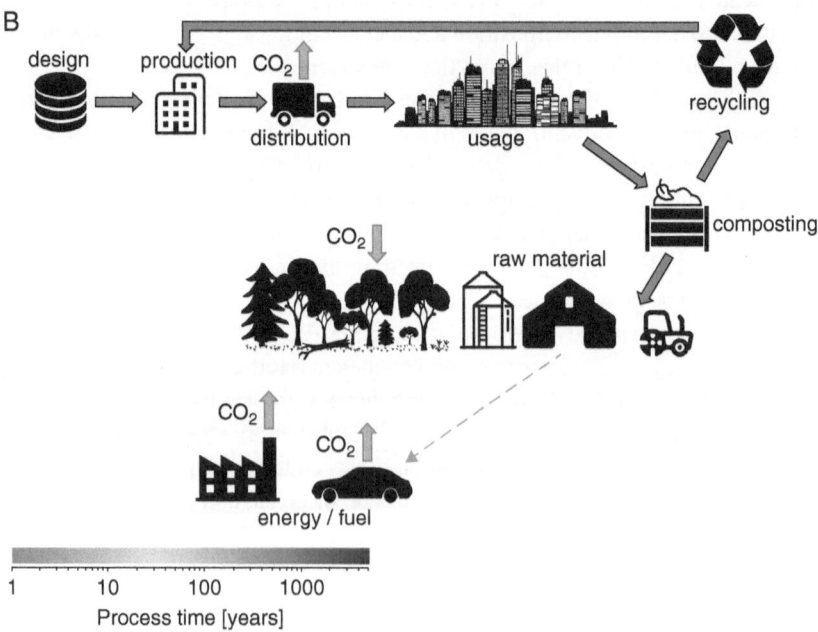

FIGURE 12.1 Carbon life cycle.

and Germany, virtually no plastic waste is being disposed of in landfills. In these countries, over 70% of plastic waste is managed through energy recovery, while the rest is materially recycled. Some European Union (EU) member states, including Poland, Spain, and Greece, trail far behind, with less than 40% of their plastic waste being recovered (Figure 12.2). Out of 57.2 million tons of plastic produced in EU countries in 2022 over 17% underwent recycling (PlasticEurope: Plastics – The Facts 2022, 2024). Post-consumer recycled plastics and bio-based/bio-attributed plastics accounted for 10.1% and 2.3% of European plastics production, respectively.

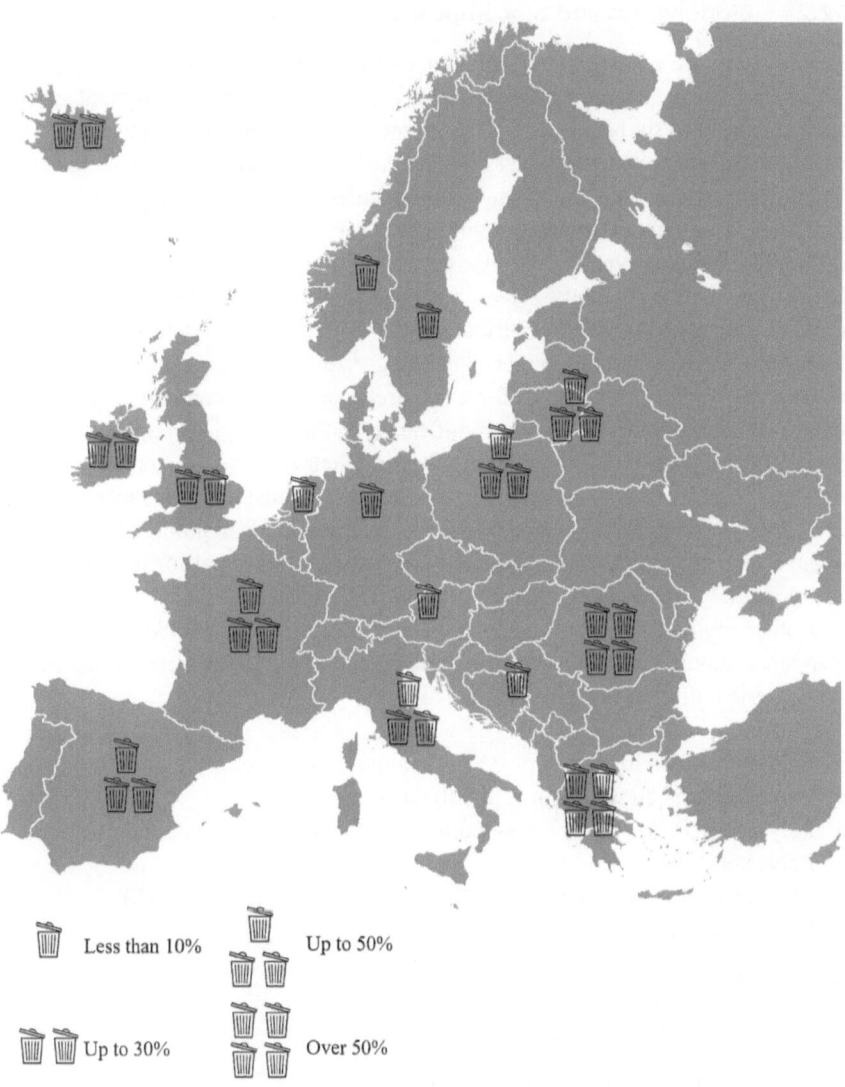

FIGURE12.2   Plastics post-consumer waste landfill rate across Europe.

The adoption of biopolymers by means of additive manufacturing techniques (Mehrpouya et al., 2021) will also play a decisive role in this transition toward more extensive use of biopolymers. The 3D printing materials market recorded a turnover of 2,578.8 million dollars in 2022 and is expected to increase by 25.9%, reaching 16,230.8 million dollars by 2030, due to the increasingly massive use of additive manufacturing in production for mass customization (*3D Printing High-Performance Plastics Market Research Report*, 2023).

12.2   Biopolymers and biocomposites

Biopolymers are a category of polymer that is synthesized by living organisms. This diverse group encompasses a range of sources, including various plant species, such as corn and soybeans, as well as certain trees and specific bacteria (Vroman et al., 2009). These polymers, classified on the basis of their structural composition and the monomeric units that form their molecules, are biomolecules that play integral roles in biological processes (Witko, 2019; Witko et al., 2019). The formation of biopolymers involves the intricate orchestration of biological systems, wherein living organisms meticulously assemble these complex molecules. The monomeric units utilized in this process contribute to the distinct characteristics and functionalities of the resulting biopolymer. This categorization based on both structure and monomeric composition allows for a comprehensive understanding of the diverse nature of biopolymers and their significance in the biological realm. Moreover, the applications of biopolymers extend beyond their role in living organisms. Harnessing the properties of biopolymers has become a focal point in various sectors, including in biotechnology and materials science. Understanding the structural nuances and monomeric foundations of biopolymers is crucial for unlocking their full potential in the development of sustainable materials, medical applications, and other innovative technologies (Malagurski et al., 2021; Witko et al., 2018). Polymeric biomolecules can be classified depending on the polymer structure and monomeric unit used to form the molecule:

- Polynucleotides like DNA or RNA – composed of 13 or more monomers
- Polypeptides – short amino acid polymers
- Polysaccharides – linear carbohydrate structures depending on their origin
- Polyesters (PHAs, PLA)
- Proteins (Silk, Collagen, Gelatin Soy, Zein, Gluten, Albumin, etc.)
- Polysaccharides:

  - Bacterial (Xanthan, Dextran)
  - Fungal (Pollulan, Elsinan)
  - Algal, Plant (Starch, Cellulose)
  - Animal (Chitin, Hyaluronic Acid)

- Lipids (Acetoglycerides, Waxes)
- Polyphenols (Lignin, Tannin)
- Other polymers (natural rubber, PGA(poly(glycolic acid), polymers from fats and oils)

The term biodegradation is associated with materials that can be broken down by natural processes either through chemical mechanisms without enzyme contribution (hydrolysis) or enzymatic-driven mechanisms (Velema et al., 2006). This particular property of material depends on many factors, and one of them is the origin of the polymer (Figure 12.3).

The inception of man-made bioplastics can be traced back to Alexander Parkes, who unveiled the first bioplastic of its kind, known as Parkesine, at the 1862

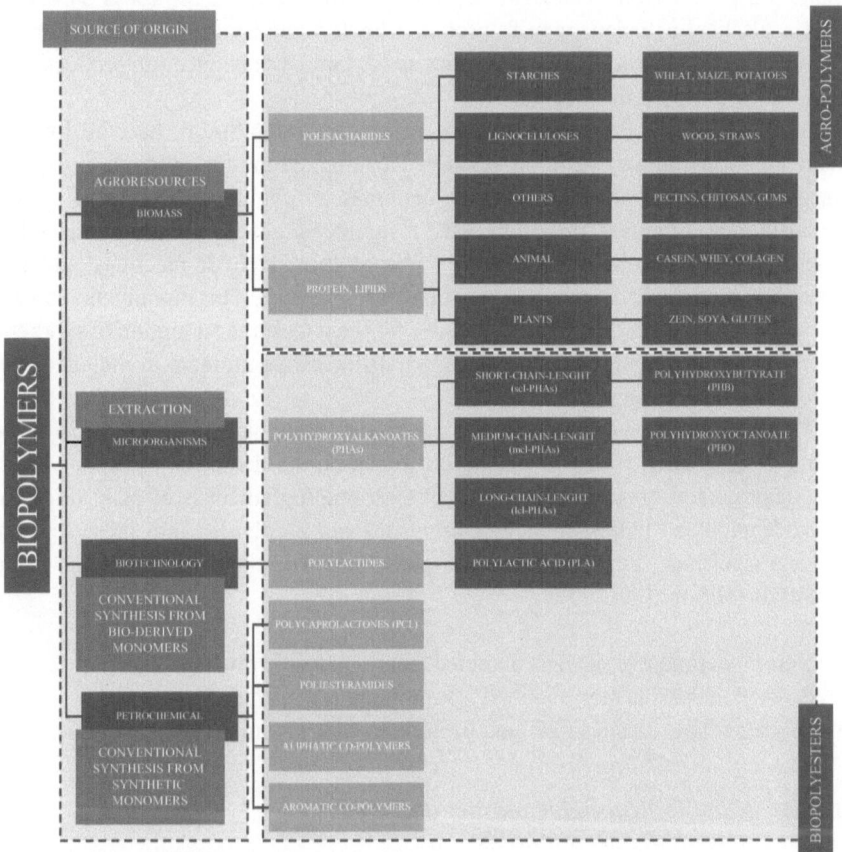

FIGURE 12.3   Biopolymer classification by source of origin, petroleum-based biodegradable polymers are included on this graph.

Great International Exhibition in London. This groundbreaking material marked a significant leap in biomaterial innovation, as it was derived from cellulose, signifying a departure from traditional synthetic polymers.

Shortly thereafter, in 1868, John Wesley Hyatt introduced Celluloid, a biomaterial originating from cellulose and alcoholized camphor. Hyatt was motivated by the desire to provide an alternative to ivory for use in billiard balls. The initial cellulose polymer lacked sufficient strength until the addition of camphor, a by-product of the laurel tree. Celluloid found widespread application as the pioneering material for the first flexible photographic film, thereby revolutionizing the field of cinematography. Hyatt made another contribution to the film industry by producing celluloid in strip format, catering specifically to movie film applications.

The evolution of bioplastics continued its trajectory with formaldehyde, which marked a pivotal point in plastic technology. Around 1897, the focus shifted to casein plastics, a fusion of milk protein with formaldehyde. This technological advance led to the production of Galalith and Erinoid, two pioneering products that found their place on the market.

In 1899, British Patent no. 16,275, issued to Arthur Smith, laid the foundations for formaldehyde resin processing. This marked the first attempt to explore the potential of phenol-formaldehyde resins as an ebonite substitute in electrical insulation. The year 1907 witnessed a significant advance when Leo Hendrik Baekeland elevated the performance of phenol–formaldehyde reactions, and as a result invented the first entirely synthetic resin – Bakelite. This monumental breakthrough in resin technology paved the way for the widespread adoption of synthetic resins in various applications, marking a transformative moment in the history of bioplastics (Rasmussen, 2021).

Many different ways of classifying biopolymers/biocomposites exist. The above classification concerns origin, but based on the definition of biopolymers/biocomposites as "materials in which at least one ingredient is of plant origin or is biodegradable" biopolymers/biocomposites can be divided into those that are biodegradable and those that are not (Figure 12.4). The classification of this group is thus as follows:

- Non-biodegradable plastics obtained from renewable raw materials;
- Biodegradable plastics obtained from mining raw materials;
- Biodegradable plastics obtained from renewable raw materials.

12.2.1 Non-biodegradable plastics obtained from renewable raw materials

The first group included in the classification of biobased/biodegradable plastics comprises non-biodegradable plastics obtained from renewable raw materials such as corn, sugar cane, and biomass. The production process includes pre-treatment,

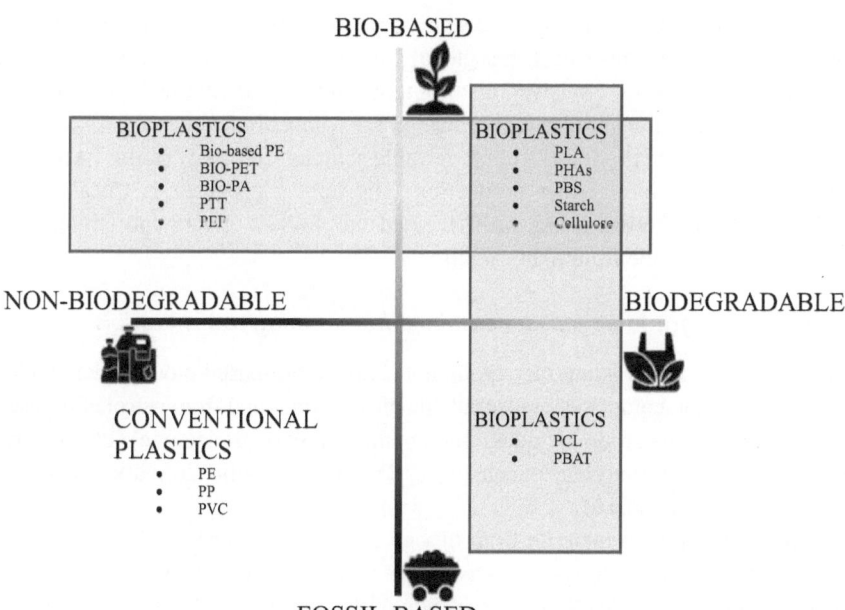

BIO-BASED

BIOPLASTICS
- · Bio-based PE
- BIO-PET
- BIO-PA
- PTT
- PEF

BIOPLASTICS
- PLA
- PHAs
- PBS
- Starch
- Cellulose

NON-BIODEGRADABLE   **BIODEGRADABLE**

CONVENTIONAL
PLASTICS
- PE
- PP
- PVC

BIOPLASTICS
- PCL
- PBAT

FOSSIL-BASED

FIGURE 12.4   Classification of biopolymers according to their biodegradability and the carbon source from which they were produced.

hydrolysis, fermentation, and several organic reaction stages. One such example is polyamide 11 (PA11), which is 100% bio-based. This is a semi-crystalline polymer obtained from castor oil. Due to its renewable sources of production, this material is a good alternative to petroleum-based polyamide (PA6 and PA6.6). PA11, in addition to its good mechanical properties compared to PA, has a low specific gravity (~1.03 g/cm^3), excellent chemical resistance, impact and abrasion resistance, a relatively high melting point ($T_m = \sim180°C–190°C$), and a wide processing temperature range from 130°C to 240°C (Martino et al., 2014). Moreover, its functional end groups and amide bonds enable the formation of hydrogen bonds and good interfacial interaction with natural fibers in the production of composites (Jacques et al., 2002).

Another biocompatible plastic that originates from nature but non-biodegradable plastic is bio-based polyethylene (bioPE), which has been gaining in popularity recently as a replacement for its petroleum-based counterpart, one of the most commonly used plastics in the world. BioPE accounted for 10.5% of the biopolymer market in 2019, behind only PLA (18.7%), starch blends (18.7%), PBAT (13.5%), and bioPA (11.9%) (Seculi et al., 2023). The basic substrate of PE is ethanol. For the bio-based version sugar cane is used (produced mainly in Brazil), although this plastic can be produced using any raw material suitable for ethanol production (Suarez et al., 2023).

PET is the most commonly used semi-aromatic polyester. It is the fourth most produced plastic in the world, the global supply of which in 2012 amounted to over 19.8 million tons. However, disturbing information about the depletion of oil deposits provided the impulse for replacing PET with a biopolymer. PET's competitor is its equivalent produced from renewable sources – bioPET. BioPET accounts for 26% of total biopolymer production. The largest producer is Coca-Cola®, which makes its bottles using bioPET based on ethylene glycol and terephthalic acid, which are made from plant sugars.

12.2.2 Biodegradable plastics obtained from mining raw materials

The second, and least numerous, group in the above bio-based/biodegradable plastics classification comprises biodegradable plastics obtained from petroleum-based raw materials. This group includes, among others: poly(caprolactone) (PCL), polyvinyl alcohol, poly(butylene succinate) (PBS), and aliphatic–aromatic polyesters (AACs) (A. Prieto, 2016).

The most common material from this group is PCL, which has been produced since the mid-1970s. PCL is a semi-crystalline and linear aliphatic polyester obtained by means of chemical synthesis from crude oil via the ring-opening polymerization of caprolactone monomer. This material has good mechanical properties and it is widely used in industry, mainly in medicine due to its biocompatible properties and biodegradability. PCL degradation occurs in two stages: (i) random cleavage of the hydrolytic ester and (ii) mass loss through the diffusion of oligomeric species from the mass. The biodegradation process then depends on the bioavailability of water, which promotes microbial attack and matrix hydrolysis (Oztemur et al., 2021).

One biodegradable petrochemical polymer worth noting is PBS. PBS is obtained by reacting diacid or acid anhydride with diols and with water eliminated (Aliotta et al., 2022). Until recently, this material could only be obtained from petrochemical raw materials, but nowadays its synthesis from biobased materials is also possible. It possesses good physical and mechanical properties similar to PP or PE, but it is also characterized by high chemical and thermal resistance and its processing is simple (temperature range 160°C–200°C). The main application of PBS is in agricultural products (padded parts, collars) and the packaging industry (disposable cutlery, household, bottles, etc.).

Besides PCL another interesting aliphatic polyester is AAC, which consists of aliphatic and aromatic units, in particular poly(butylene adipate coterephthalate) (PBAT). PBAT synthesis occurs through the polycondensation of diols and dicarboxylic acids – butanediol, adipic acid, and terephthalic acid (Wu, 2012).

PBAT is not only characterized by good biodegradability but also due to the aliphatic segment in its polymer structure, it also possesses good mechanical properties. Compared to most biodegradable polyesters, such as PLA and PBS, the mechanical properties of PBAT are characterized by greater flexibility and are similar to those of low-density PE (Yin et al., 2019).

12.2.3 Biodegradable plastics obtained from renewable raw materials

One of the most common biodegradable plastics is PLA. It was first synthesized in 1932, and then launched on the market in 1954, when it was used in medicine and tissue engineering. It is obtained from natural raw materials such as corn, sugar cane, and sugar beet (Bergström et al., 2016). It is obtained by fermenting starch, from which lactic acid is obtained and PLA is synthesized on its basis. The synthesis of PLA is a multi-stage process involving polymerization with the opening of the latite chain. PLA owes its biodegradability to the hydroxyl and carboxyl groups, thanks to which it is easily hydrolyzed (Bergström et al., 2016; Hyon et al., 1997).

Another group of biodegradable plastics consists of PHA polymers, first synthesized in 1927. PHAs are aliphatic polyesters produced by microorganisms that undergo a process of fermentation in order to accumulate a carbon source and energy storage material.

One of the most popular polymers from this group is poly(3-hydroxybutyrate -*co*-3-hydroxyvalerate) (PHBV) (Bledzki et al., 2010; Kuciel et al., 2019). It is a copolymer with poly(3-hydroxybutyrate) and hydroxyvalerate (HV). The physical and mechanical properties of PHBV are similar to those of conventional thermoplastics such as PE and PP, but it is biodegradable and biocompatible, and its properties can easily be altered by changing the HV content. A low content of HV in PHBV leads to a decrease in elasticity, elongation, and impact strength, while a higher content reduces its brittleness due to the inhibition of secondary crystallization (Mohanty et al., 2002).

12.3 Biomaterials

For more than half a century, the exploration of biopolymers was confined to the realms of biochemistry and molecular biology. However, in the past decade, a paradigm shift has occurred, and researchers in soft matter physics, biophysics, chemistry, and materials science have eagerly embraced this field of study. The contemporary surge in interest in developing novel materials has propelled a demand for biological and medical studies, especially for various clinical applications. The concept of biomaterials has emerged as a pivotal framework, driven by the need to integrate materials science into the medical field (Langer et al., 2004).

Defined as materials used to create devices that replace a part or function of an organism in a safe, reliable, economic, and physiologically acceptable manner, biomaterials necessitate interdisciplinary collaboration to ensure they are developed and utilized in bioengineering or medicine efficiently and effectively. The historical selection of materials for medical implants was initially limited to already known substances, such as metals like gold, which was first used in dentistry over 2,000 years ago (Yadav et al., 2015).

The term "biomaterials" was officially introduced in the 1950s, and thus coincided with the industrial expansion of polymer synthesis. Synthetic polymers, such as polymethylmethacrylate and cellulose acetate, began to be employed for biomedical purposes in the 1930s and 1940s, respectively. While naturally occurring materials like collagen were also utilized, materials were often adopted from other scientific and technological domains for medical use without any major redesigning (Khan et al., 2013).

Early implants and medical devices, derived from industrially used materials, frequently proved incompatible with host tissues, leading to such problems as recoils and a frequent failure to adequately perform their intended functions. Research areas were consequently established to explore polymer features for biomedical applications and carry out biocompatibility testing. Subsequent advances in immunology and a better understanding of foreign body reactions led to the development of first-generation biomaterials in the 1960s and 1970s. These materials aimed to match the mechanical properties of replaced tissues and reduce immune responses and rejections.

The early 1980s marked a significant shift in research focus, transitioning from bio-inert materials to substances actively interacting with their environment. Second-generation biomaterials were designed to be bioactive, eliciting specific cellular responses such as adhesion, proliferation, and differentiation. Biodegradability, allowing controlled chemical disintegration into non-toxic products, represented a pivotal advance. The third generation of biomaterials, which emerged later, integrates both resorbable and bioactive properties, and their purpose was to facilitate the body's self-healing mechanisms (Drotleff et al., 2004).

It is worth noting that the current trend is toward to a greater share of biopolymers in overall biomaterials production. These eco-friendly alternatives are steadily replacing traditional petroleum-based polymers in various industrial applications, signaling a transformative shift toward sustainable materials.

12.4 Sustainable product innovations based on the example of biopolymers

In recent years, global awareness of the environmental impact of traditional plastics has increased, leading to a surge in sustainable product innovations (Baranwal et al., 2022). Among these innovations, biopolymers have emerged as a promising alternative, offering an eco-friendlier option for various industries (Skibiński et al., 2021). This chapter explores the concept of sustainable product innovations with a focus on biopolymers, discussing their properties, current and future applications, as well as their potential for promoting a more sustainable future.

Biopolymers showcase a diverse set of characteristics that render them an appealing choice for innovations in sustainable products. Their biodegradability, that is, their ability to naturally break down into harmless by-products, contributes to the goal of mitigating the environmental impact of plastic waste. Moreover, biopolymers frequently demonstrate mechanical properties comparable with those of traditional polymers and composites, rendering them well-suited to a variety of applications.

The adaptability of biopolymers has resulted in their widespread application in diverse industries (Babu et al., 2013; Baranwal et al., 2022; Koller et al., 2018). For instance, in the area of packaging, biodegradable film derived from starch-based polymers is being increasingly used as a more environmentally friendly substitute for traditional plastic packaging. Simultaneously, the textile industry has embraced such biopolymers as PLA acid in the manufacture sustainable fabrics and garments, which is in line with increasing demand for eco-conscious fashion.

A particularly noteworthy application of biopolymers has been observed in the biomedical sector, where their growing prominence stems from such inherent qualities as biocompatibility and degradability thanks to their being harmlessly absorbed (and then excreted) by the body (Witko et al., 2019). Biopolymers have emerged as vital materials used in medical implants, drug delivery systems, and scaffolds for tissue engineering. The dual functionality of biopolymers, that is, not only offering benefits in terms of sustainability but also an ability to seamlessly integrate with biological systems, underscores their potential for catalyzing trans-formative changes across a spectrum of industries.

As advances in biopolymer research continue, the biomedical field stands poised to make groundbreaking innovations. The biodegradability and compatibil-ity of biopolymers position them as instrumental components in the quest for more sustainable and biologically harmonious solutions, further highlighting their role as a lever for a positive change in numerous industrial applications (Malagurski et al., 2021; Witko et al., 2020).

12.5   Conclusions

12.5.1   Packaging industry and food industry

The packaging sector, which is a major contributor to environmental problems due to its extensive utilization of conventional plastics, is undergoing a signifi-cant transformation thanks to the incorporation of biopolymers (Amass et al., 1998). Sourced from renewable materials, biopolymers represent an environ-mentally conscious substitute for traditional packaging components. This chapter investigates the present state of biopolymers in the packaging industry, providing instances of particular products that have showcased their positive and sustain-able influence. Numerous types of biopolymer are currently being used in this branch of industry.

Starch-based biopolymers have become an increasingly popular sustainable choice in packaging. These polymers are often derived from corn, potatoes, or other starch-rich crops. Biodegradable films made from starch-based polymers are increasingly utilized in various packaging applications. Notable examples include compostable bags, food packaging, and disposable tableware. The introduction (also thanks to new legislature introduced by some European countries) of biodegradable shopping bags made from starch-based biopolymers marks a significant stride toward sustainability, offering a viable substitute for conventional plastic bags with a considerable environmental impact. These innovative bags undergo natural decomposition, thereby contributing in a noteworthy way to the task of mitigating the environmental consequences of single-use plastics. By embracing this eco-friendly alternative, we not only address the concerns of plastic pollution but also promote a more sustainable and environmentally responsible approach to daily consumer practices. Biopolymer films, particularly those made from starch-based polymers, are employed as eco-friendly wraps for fresh produce. These wraps offer protection while decomposing naturally, minimizing the environmental footprint associated with conventional plastic wraps.

PLA, a biopolymer derived from fermented plant sugars, is widely employed in packaging applications. Its versatility allows for the production of clear, rigid, and flexible packaging materials. PLA-based products include clear cups, food containers, and packaging films, thereby offering a sustainable alternative with the added benefit of being compostable in the right conditions. PLA-based biopolymers play a pivotal role in the manufacturing of compostable food containers, which are particularly prevalent in takeout packaging. These containers serve as an eco-conscious solution for the foodservice sector, presenting packaging that aligns with sustainability goals by allowing for composting after use. This environmentally friendly approach not only addresses the challenges associated with traditional plastic waste but also helps foster a more sustainable and responsible food packaging industry. As consumer awareness continues to grow, the utilization of PLA-based biopolymers in compostable containers stands as a testament to the industry's commitment to reducing its environmental footprint.

PHA-based packaging materials offer excellent biodegradability and can be used in various applications, including packaging for perishable goods (Amass et al., 1998; Tharanathan, 2003). PHAs stand out as valuable biopolymers, showcasing significant promise in revolutionizing the packaging industry. Leveraging their inherent biodegradability, PHAs are harnessed to craft packaging solutions that are both sustainable and environmentally friendly. In providing an alternative to conventional plastics, which pose substantial environmental concerns, PHA-based packaging materials offer a proactive response to the pressing issue of plastic pollution. The versatility of PHAs extends across various packaging applications, playing a crucial role in the development of biodegradable films, bags, and containers. Their innate capacity to decompose naturally not only alleviates environmental concerns but also positions PHAs as a compelling eco-friendly choice for both consumers and industries. As the demand for sustainable packaging solutions surges, the integration of PHAs into the packaging sector signifies a commendable and significant step toward mitigating the environmental impact associated with the single-use plastics (Follain et al., 2014).

Continuous research and development endeavors in this domain are focused on elevating the performance and cost-effectiveness of PHA-based packaging. These efforts aim to foster a more sustainable and circular approach to packaging materials, reflecting a commitment to environmental stewardship in the packaging industry. The trajectory of PHAs in packaging underscores their potential to reshape the landscape, ushering in an era where ecological responsibility harmonizes with packaging innovation.

In addition, biopolymers feature in encapsulation systems for food ingredients and additives. This solution enables the controlled release and improved stability of sensitive components, contributing in this way to the overall quality and safety of food products. Overall, the integration of biopolymers into the food industry accords with the growing demand for sustainable practices, providing innovative solutions aimed at reducing the environmental footprint of packaging and raising the overall quality of food products.

Starch-based biopolymers can be processed into films and coatings, providing an eco-friendly alternative to traditional petroleum-based packaging. For example, starch-based biopolymers have been utilized to create biodegradable trays for fresh products, thereby countering the negative environmental impact of single-use plastic packaging.

Additionally, starch-based biopolymers play a crucial role in the development of edible film. These films, made from starch and other natural polymers, can serve as coatings for fruit or as wrappers for food items, extending their shelf life while remaining safe for consumption.

PLA and PHAs can be utilized to create biodegradable films and containers. For instance, PLA-based clamshell containers and cups have become popular choices for serving food and beverages due to their eco-friendly nature (Solarz et al., 2023).

PLA and PHAs are also employed in the production of biodegradable utensils, such as cutlery and straws, thus helping reduce single-use plastic waste in the food service sector. Besides packaging, PLA is also a component in transparent lids and windows for food containers, providing visibility to consumers while maintaining the material's biodegradable properties.

PLA and PHAs have ventured beyond packaging into the realm of edible films. These films, made from PLA, can be used as coatings for fruit and vegetables, enhancing shelf life and reducing food waste.

12.5.2 Textile industry

The textile industry, a traditionally resource-intensive sector, is undergoing a significant transformation driven by the adoption of sustainable materials (Olkhov et al., 2015). Derived from renewable resources biopolymers have emerged as key players in this paradigm shift. As a consequence PLA, PHA and starch-based polymers have gained traction in the textile industry due to their eco-friendly features. PLA, derived from plant-based sources like corn or sugarcane, is a notable example. Its biodegradability and ability to be processed into fibers make it an attractive alternative to conventional synthetic materials.

Starch-based biopolymers derived from renewable resources, such as corn, wheat, or potatoes, offer a sustainable solution due to their biodegradability and lower environmental impact. One of their notable applications is in biodegradable packaging for textiles, where starch-based polymers help minimize the industry's reliance on conventional plastic packaging. Moreover, starch-based biopolymers are utilized in the production of eco-friendly fabrics. Fibers derived from starch can be processed to create textiles that exhibit desirable qualities such as breathability and comfort. Some forward-thinking clothing brands are incorporating these starch-based textiles into their collections, providing consumers with sustainable fashion choices.

In addition to textiles, starch-based biopolymers have other applications, for example in accessories such as hats, bags, and footwear. These biodegradable alternatives offer a viable option for environmentally conscious consumers seeking sustainable products for various aspects of their daily lives. They are likewise used in goods with short lifespans. The versatility and eco-friendly nature of starch-based biopolymers showcase their potential to revolutionize the textile industry and contribute to a more sustainable and circular economy.

Biopolymers, especially PLA, are used in the production of eco-friendly fabrics. These fabrics are present in various clothing items, from casual wear to sportswear. Brands committed to sustainable practices incorporate these textiles into their collections, providing consumers with environmentally conscious choices.

Biopolymers have found their way into the production of footwear and accessories. Shoes made from biodegradable materials offer a sustainable option for consumers seeking environmentally conscious choices in their fashion and lifestyle products (Younes, 2017).

PHAs have emerged as valuable biopolymers with diverse applications, and the textile industry is increasingly exploring their potential. One notable application of PHAs in the textile sector is in sustainable fibers. Being biodegradable and derived from renewable sources PHAs, offer an eco-friendly alternative to traditional synthetic fibers. These biopolymers can be processed into fibers suitable for textile manufacturing, addressing environmental concerns associated with non-biodegradable materials. PHAs also play an important role in the creation of biodegradable textiles and fabrics, thus aligning with the industry's growing emphasis on sustainability. As the textile industry continues to seek innovative and sustainable solutions, the versatile properties of PHAs position them as promising materials for shaping a more eco-friendly and socially responsible future in textile production.

12.5.3 Agriculture

Biopolymers have had a major impact on agriculture, helping promote sustainable and environmentally friendly practices. One notable example is the use of biodegradable mulching film made from starch-based biopolymers. This type of film has replaced traditional plastic mulch, leading to improved yields and crop traits that avoid heavy dependence on agrochemicals. They also provide weed control and moisture retention benefits while naturally breaking down over time, thus minimizing their environmental impact (Menossi et al., 2021). Additionally, biopolymers are employed in soil erosion control. Biodegradable erosion control blankets made from natural polymers assist in preventing soil erosion, promoting vegetation growth, and ultimately enhancing soil stability. These applications showcase the versatility of biopolymers in addressing critical agricultural challenges while aligning with the global push toward more sustainable farming practices. Furthermore, biopolymers have been utilized in seed coatings to enhance germination and protect seeds from environmental stresses. By encapsulating seeds in biodegradable polymer coatings, the release of nutrients can be controlled, providing a favorable environment for seedlings to thrive. Bio-polymers have replaced synthetic alternatives in the formulation of hydrogels, since they exhibit good biodegradability and biocompatibility. Natural hydrogels, based on starch, chitosan, guar gum, alginate, lignin, and gelatin, have broad agricultural application due to their ability to retain water and tune the release of fertilizers and/or agrochemicals. As a consequence, biopolymer-based hydrogels, for which further research is envisaged aimed at improving their mechanical strength and reducing production costs (Chen, 2020), represent a new frontier in agriculture in the field of soil conditioners and nutrient reservoirs (Gutiérrez, 2019). These innovative applications demonstrate the potential of biopolymers to revolutionize various aspects of agriculture, fostering a more sustainable and eco-friendly approach to farming (Cinelli et al., 2019; Juturu et al., 2016).

12.5.4 Medicine and biomedical engineering

Biopolymers stand out as excellent materials for medical applications due to a combination of unique properties that address critical needs in healthcare (Feliksiak et al., 2020). Firstly, their biocompatibility ensures minimal adverse reactions when interacting with biological systems, thereby reducing the risk of inflammation or rejection. Secondly, thanks to their biodegradability these materials break down naturally over time, eliminating the need for surgical removal and reducing long-term complications. Additionally, biopolymers often exhibit low levels of immunogenicity, meaning they provoke a minimal immune response, further enhancing their suitability for medical use (Witko, 2019; Witko et al., 2020). Biopolymers offer versatility in fabrication, allowing for the creation of complex structures tailored to specific medical needs. Their tunable mechanical properties enable the design of materials ranging from soft hydrogels for tissue engineering to rigid implants for orthopedic applications. Furthermore, many biopolymers possess inherent antimicrobial properties, aiding in the prevention of infections, which is crucial in medical settings. The unique, circular nature of biopolymers, sourced from renewable biomass such as plants or microorganisms, aligns with sustainable practices, reducing dependency on fossil fuels and minimizing environmental impact. Such sustainability is increasingly valued in the medical field, where reducing waste and environmental footprint is a growing concern. The combination of biocompatibility, biodegradability, tunable properties, antimicrobial characteristics, and sustainability makes biopolymers exceptional materials for a wide range of medical applications, promising safer and more effective treatments for patients while promoting environmental responsibility in healthcare. Nowadays, biopolymers are present in the biomedical industry, and are widely used in many types of therapies. Biopolymer-based drug delivery systems offer controlled release of therapeutic agents, improving efficacy and fewer side effects. For instance, hydrogels composed of alginate or chitosan can encapsulate drugs and release them at a controlled rate, enhancing targeted delivery to specific tissues. Biopolymeric materials can serve as scaffolds in tissue engineering to support cell growth and tissue regeneration. Materials such as collagen, gelatin, and hyaluronic acid mimic the extracellular matrix, promoting cell adhesion, and proliferation. These scaffolds are used to repair damaged tissue or organs, such as in skin grafts, or promote cartilage regeneration. Bio-based sutures made from polymers such as polyglycolic acid or PLA provide excellent tensile strength and biodegradability, thereby reducing the risk of inflammation and promoting wound healing. Biopolymer-based wound dressings, such as those incorporating chitosan or alginate, offer antimicrobial properties and maintain a moist wound environment conducive to healing. Biopolymers are increasingly utilized in the fabrication of biodegradable implants for temporary structural support or drug delivery. Poly(lactic-*co*-glycolic acid) is a commonly used biopolymer for orthopedic implants and surgical screws, gradually degrading in the body without causing harm. In addition, the ability of materials like PLA to be processed as filaments in fused filament fabrication allow material engineers to design particular prosthesis shapes for body part replacement. Some of these materials can also be included in diagnostic tools such as biosensors and bioimaging agents. For instance, protein-based biopolymers like antibodies are used in biosensors for detecting specific biomarkers indicative of certain diseases, thus offering rapid and accurate diagnoses. Biopolymer-based vectors are utilized in gene therapy to deliver therapeutic genes into target cells. Polymers like polyethyleneimine or chitosan form complexes with nucleic acids, protecting them from degradation and facilitating cellular uptake, thereby enhancing gene delivery efficiency.

12.5.5 Future applications

Future applications of biopolymers hold immense promise in addressing global environmental challenges and fostering sustainable development across various industries. Derived from renewable resources and boasting biodegradability, biopolymers are gaining momentum as viable alternatives to traditional petroleum-based plastics (Rai et al., 2011; Vandi et al., 2018). This transition fits into the broader context characterized by the phasing out of fossil fuels as a source of energy and the desire to away from crude oil as a source of platform chemicals for polymer synthesis. In light of efforts to meet the United Nations' Sustainable Development Goals, when it comes to fossil feedstock replacement biorefineries have emerged as an alternative to traditional refineries for producing chemical building blocks, and materials out of biomass (Calvo-Flores et al., 2022). Biopolymers' versatile properties pave the way for a myriad of potential applications across various sectors, promising innovative solutions to pressing environmental challenges. In the future, biopolymers may serve as scaffolds for cell growth and regeneration. Another area in which they perform a valuable function is in drug delivery systems, where their controlled degradation allows for the sustained release of pharmaceuticals. Future advances may see the integration of these biopolymers into implantable medical devices and surgical materials, offering safer and more sustainable healthcare solutions. In the construction sector, biopolymers hold enormous potential in helping develop eco-friendly building materials. Biodegradable polymers can be incorporated into composites, insulation materials, and coatings, reducing reliance on fossil fuel-derived resources and lowering carbon emissions. In the future, biopolymers may be used in modular construction systems, green roofs, and sustainable infrastructure projects, thus supporting the construction of energy-efficient and environmentally conscious buildings. The versatility of biopolymers extends to the realm of 3D printing and additive manufacturing. PLA, in particular, has become increasingly attractive as a bio-based filament for 3D printing due to the ease of its processing and its biodegradability (Giubilini et al., 2021). Future developments may lead to the synthesis of specialized biopolymers tailored to additive manufacturing, offering sustainable alternatives to conventional thermoplastics. This could revolutionize the production of custom-designed parts, prototypes, and functional objects with minimal environmental impact. Biopolymers may also play a vital role in advancing renewable energy technologies (Verlinden et al., 2007; Dalwadi et al., 2023; Gopinath et al., 2023; Joshi et al., 2024). From biodegradable components for solar panels to bio-based materials for energy storage devices, their applications are diverse and promising (Hasan et al., 2020; Mohiuddin et al., 2017). In the future, the unique properties of biopolymers may be harnessed for the purposes of enhancing the efficiency and sustainability of renewable energy systems, as well as to facilitate the transition toward a greener energy landscape.

The primary driving forces behind the global biopolymer market are environmental conservation and the pressing need to reduce reliance on fossil fuels. Both the European and global biopolymer markets enjoy consistent growth, propelled by increasing regulatory restrictions placed on the use of conventional plastics in specific applications. PHAs, among other biopolymers, possess desirable physical properties such as a high Young's modulus and thermoplastic characteristics, making them suitable for modern manufacturing techniques like 3D printing or electrospinning. The adoption of biopolymers in various industries and medical fields not only promotes environmental protection and aids in CO_2 emission reduction but also accelerates advances in technology and the biomedical sciences. Previously, material limitations hindered the progress of certain technologies and impeded effective medical therapies, but biopolymers offer solutions to these challenges.

Acknowledgement

The publication/article presents the result of the Project no 070/ZJE/2024/POT financed from the subsidy granted to the Krakow University of Economics.

References

3D Printing High-Performance Plastics Market Research Report (2023). Delhi. Retrieved from https://www.psmarketresearch.com/market-analysis/3d-printing-high-performance-plastics-market

Aliotta, L., Seggiani, M., Lazzeri, A., Gigante, V., & Cinelli, P. (2022). A Brief Review of Poly (Butylene Succinate) (PBS) and Its Main Copolymers: Synthesis, Blends, Composites, Biodegradability, and Applications. *Polymers, 14*(4), 844. doi: 10.3390/POLYM14040844

Amass, W., Amass, A., & Tighe, B. (1998). Review of Biodegradable Polymers: Uses, Current Developments in the Synthesis and Characterization of Biodegradable Polyesters, Blends of Biodegradable Polymers and Recent Advances in Biodegradation Studies. *Polymer International, 47*, 89–144. doi: 10.1002/(SICI)1097-0126(1998100)47:2<89::AID-PI86>3.0.CO;2-F

Babu, R. P., Connor, K. O., Seeram, R., O'Connor, K., & Seeram, R. (2013). Current Progress on Bio-Based Polymers and Their Future Trends. *Progress in Biomaterials, 2*(1). doi: 10.1186/2194-0517-2-8

Baranwal, J., Barse, B., Fais, A., Delogu, G. L., & Kumar, A. (2022). Biopolymer: A Sustainable Material for Food and Medical Applications. *Polymers, 14*(5), 983. doi: 10.3390/polym14050983

Bergström, J. S., & Hayman, D. (2016). An Overview of Mechanical Properties and Material Modeling of Polylactide (PLA) for Medical Applications. *Annals of Biomedical Engineering, 44*(2), 330–340. doi: 10.1007/s10439-015-1455-8

Bledzki, A. K., & Jaszkiewicz, A. (2010). Mechanical Performance of Biocomposites Based on PLA and PHBV Reinforced with Natural Fibres – A Comparative Study to PP. *Composites Science and Technology.* doi: 10.1016/j.compscitech.2010.06.005

Calvo-Flores, F. G., & Martin-Martinez, F. J. (2022). Biorefineries: Achievements and Challenges for a Bio-Based Economy. *Frontiers in Chemistry, 10.* doi: 10.3389/fchem.2022.973417

Chakrabarty, G., Vashishtha, M., & Leeder, D. (2015). Polyethylene in Knee Arthroplasty: A Review. *Journal of Clinical Orthopaedics and Trauma, 6*(2), 108–112. doi: 10.1016/J.JCOT.2015.01.096

Chen, Y. (Ed.) (2020). *Hydrogels Based on Natural Polymers.* Elsevier. doi: 10.1016/C2018-0-00171-1

Cinelli, P., Seggiani, M., Mallegni, N., Gigante, V., & Lazzeri, A. (2019). Processability and Degradability of PHA-Based Composites in Terrestrial Environments. *International Journal of Molecular Sciences*, *20*(2), 284. doi: 10.3390/ijms20020284

Dalwadi, S., Goel, A., Kapetanakis, C., Salas-de la Cruz, D., & Hu, X. (2023). The Integration of Biopolymer-Based Materials for Energy Storage Applications: A Review. *International Journal of Molecular Sciences*, *24*(4), 3975. doi: 10.3390/ijms24043975

Drotleff, S., Lungwitz, U., Breunig, M., Dennis, A., Blunk, T., Tessmar, J., & Göpferich, A. (2004). Biomimetic Polymers in Pharmaceutical and Biomedical Sciences. *European Journal of Pharmaceutics and Biopharmaceutics*, *58*(2), 385–407. doi: 10.1016/j.ejpb.2004.03.018

Feliksiak, K., Witko, T., Solarz, D., Guzik, M., & Rajfur, Z. (2020). Vimentin Association with Nuclear Grooves in Normal MEF 3T3 Cells. *International Journal of Molecular Sciences*, *21*(20), 1–13. doi: 10.3390/ijms21207478

Follain, N., Chappey, C., Dargent, E., Chivrac, F., Crétois, R., & Marais, S. (2014). Structure and Barrier Properties of Biodegradable Polyhydroxyalkanoate Films. *Journal of Physical Chemistry C*, *118*(12), 6165–6177. doi: 10.1021/jp408150k

Giubilini, A., Bondioli, F., Messori, M., Nyström, G., & Siqueira, G. (2021). Advantages of Additive Manufacturing for Biomedical Applications of Polyhydroxyalkanoates. *Bioengineering*, *8*(2), 29. doi: 10.3390/bioengineering8020029

Gopinath, G., Ayyasamy, S., Shanmugaraj, P., Swaminathan, R., Subbiah, K., & Kandasamy, S. (2023). Effects of Biopolymers in Energy Storage Applications: A State-of-the-Art Review. *Journal of Energy Storage*, *70*, 108065. doi: 10.1016/j.est.2023.108065

Gutiérrez, T. J. (Ed.) (2019). *Polymers for Agri-Food Applications*. Cham: Springer International Publishing. doi: 10.1007/978-3-030-19416-1

Hasan, M. M., Islam, M. D., & Rashid, T. U. (2020). Biopolymer-Based Electrolytes for Dye-Sensitized Solar Cells: A Critical Review. *Energy & Fuels*, *34*(12), 15634–15671. doi: 10.1021/acs.energyfuels.0c03396

Hottle, T. A., Bilec, M. M., & Landis, A. E. (2017). Biopolymer Production and End of Life Comparisons Using Life Cycle Assessment. *Resources, Conservation and Recycling*, *122*, 295–306. doi: 10.1016/j.resconrec.2017.03.002

Hyon, S. H., Jamshidi, K., & Ikada, Y. (1997). Synthesis of Polylactides with Different Molecular Weights. *Biomaterials*, *18*(22), 1503–1508. doi: 10.1016/S0142-9612(97)00076-8

Jacques, B., Werth, M., Merdas, I., Thominette, F., & Verdu, J. (2002). Hydrolytic Ageing of Polyamide 11. 1. Hydrolysis Kinetics in Water. *Polymer*, *43*(24), 6439–6447. doi: 10.1016/S0032-3861(02)00583-9

Joshi, J. S., Langwald, S. V., Ehrmann, A., & Sabantina, L. (2024). Algae-Based Biopolymers for Batteries and Biofuel Applications in Comparison with Bacterial Biopolymers—A Review. *Polymers*, *16*(5), 610. doi: 10.3390/polym16050610

Juturu, V., & Wu, J. C. (2016). Microbial Production of Lactic Acid: The Latest Development. *Critical Reviews in Biotechnology*, *36*(6), 967–977. doi: 10.3109/07388551.2015.1066305

Khan, R., & Khan, M. (2013). Use of Collagen as a Biomaterial: An Update. *Journal of Indian Society of Periodontology*, *17*(4), 539. doi: 10.4103/0972-124X.118333

Koller, M., & Braunegg, G. (2018). Advanced Approaches to Produce Polyhydroxyalkanoate (PHA) Biopolyesters in a Sustainable and Economic Fashion. *The EuroBiotech Journal*, *2*(2), 89–103. doi: 10.2478/ebtj-2018-0013

Kuciel, S., Mazur, K., & Jakubowska, P. (2019). Novel Biorenewable Composites Based on Poly (3-Hydroxybutyrate-co-3-Hydroxyvalerate) with Natural Fillers. *Journal of Polymers and the Environment*, *27*(4). doi: 10.1007/s10924-019-01392-4

Laird, K. (2022). *Bioplastics Highlights: A Last Look Back at 2022*. Retrieved from https://www.plasticsindustry.org/blog/a-look-back-at-bioplastics-week-2022/

Langer, R., & Tirrell, D. A. (2004). Designing Materials for Biology and Medicine. *Nature, 428*, 487–492. doi: 10.1038/nature02388

Lüftl, S., & Visakh, P. M. (2016). Polyethylene-Based Biocomposites and Bionanocomposites: State-of-the-Art, New Challenges and Opportunities. *Polyethylene-Based Biocomposites and Bionanocomposites,* 1–41. doi: 10.1002/9781119038467.CH1

Malagurski, I., Frison, R., Maurya, A. K., Neels, A., Andjelkovic, B., Senthamaraikannan, R., Babu, R. P., O'Connor, K. E., Witko, T., Solarz, D., & Nikodinovic-Runic, J. (2021). Polyhydroxyoctanoate Films Reinforced with Titanium Dioxide Microfibers for Biomedical Application. *Materials Letters, 285*, 129100. doi: 10.1016/j.matlet.2020.129100

Martino, L., Basilissi, L., Farina, H., Ortenzi, M. A., Zini, E., Di Silvestro, G., & Scandola, M. (2014). Bio-Based Polyamide 11: Synthesis, Rheology and Solid-State Properties of Star Structures. *European Polymer Journal, 59*, 69–77. doi: 10.1016/J.EURPOLYMJ.2014.07.012

Mehrpouya, M., Vahabi, H., Barletta, M., Laheurte, P., & Langlois, V. (2021). Additive Manufacturing of Polyhydroxyalkanoates (PHAs) Biopolymers: Materials, Printing Techniques, and Applications. *Materials Science and Engineering: C, 127*, 112216. https://doi.org/10.1016/j.msec.2021.112216

Menossi, M., Cisneros, M., Alvarez, V. A., & Casalongué, C. (2021). Current and Emerging Biodegradable Mulch Films Based on Polysaccharide Bio-Composites. A Review. *Agronomy for Sustainable Development, 41*(4), 53. doi: 10.1007/s13593-021-00685-0

Mohanty, A. K., Misra, M., & Drzal, L. T. (2002). Sustainable Bio-Composites from Renewable Resources: Opportunities and Challenges in the Green Materials World. *Journal of Polymers and the Environment, 10*(1–2), 19–26. doi: 10.1023/A:1021013921916

Mohiuddin, M., Kumar, B., & Haque, S. (2017). Biopolymer Composites in Photovoltaics and Photodetectors. In *Biopolymer Composites in Electronics* (pp. 459–486). Elsevier. doi: 10.1016/B978-0-12-809261-3.00017-6

Montava-Jordà, S., Torres-Giner, S., Ferrandiz-Bou, S., Quiles-Carrillo, L., & Montanes, N. (2019). Development of Sustainable and Cost-Competitive Injection-Molded Pieces of Partially Bio-Based Polyethylene Terephthalate through the Valorization of Cotton Textile Waste. *International Journal of Molecular Sciences, 20*, 1378. doi: 10.3390/ijms20061378

Nitkiewicz, T., Wojnarowska, M., Sołtysik, M., Kaczmarski, A., Witko, T., Ingrao, C., & Guzik, M. (2020). How Sustainable Are Biopolymers? Findings from a Life Cycle Assessment of Polyhydroxyalkanoate Production from Rapeseed-Oil Derivatives. *Science of the Total Environment, 749*, 141279. doi: 10.1016/j.scitotenv.2020.141279

Olkhov, A., Staroverova, O., Iordanskii, A., & Zaikov, G. (2015). Structure and Parameters of Polyhydroxybutyrate Nanofibers. In *Engineering Textiles* (pp. 79–88). Apple Academic Press. doi: 10.1201/b18859-4

Oztemur, J., & Yalcin-Enis, I. (2021). Development of Biodegradable Webs of PLA/PCL Blends Prepared via Electrospinning: Morphological, Chemical, and Thermal Characterization. *Journal of Biomedical Materials Research Part B: Applied Biomaterials, 109*(11), 1844–1856. doi: 10.1002/jbm.b.34846

PlasticEurope: Plastics – The Facts 2022. (2024). Plastics – The Facts 2022. October.

Prieto, A. (2016). To Be, or Not to Be Biodegradable... That Is the Question for the Bio-Based Plastics. *Microbial Biotechnology, 9*(5), 652–657. doi: 10.1111/1751-7915.12393

Prieto, M. A. (2007). From Oil to Bioplastics, a Dream Come True? *Journal of Bacteriology, 189*(2), 289–290. doi: 10.1128/JB.01576-06

Rai, R., Keshavarz, T., Roether, J. A., Boccaccini, A. R., & Roy, I. (2011). Medium Chain Length Polyhydroxyalkanoates, Promising New Biomedical Materials for the Future. *Materials Science and Engineering R: Reports*, *72*(3), 29–47. doi: 10.1016/j.mser.2010.11.002

Rasmussen, S. C. (2021). From Parkesine to Celluloid: The Birth of Organic Plastics. *Angewandte Chemie International Edition*, *60*(15), 8012–8016. doi: 10.1002/anie.202015095

Seculi, F., Espinach, F. X., Julián, F., Delgado-Aguilar, M., Mutjé, P., & Tarrés, Q. (2023). Comparative Evaluation of the Stiffness of Abaca-Fiber-Reinforced Bio-Polyethylene and High Density Polyethylene Composites. *Polymers*, *15*(5), 1096. doi: 10.3390/POLYM15051096

Siracusa, V., & Blanco, I. (2020). Bio-Polyethylene (Bio-PE), Bio-Polypropylene (Bio-PP) and Bio-Poly(ethylene terephthalate) (Bio-PET): Recent Developments in Bio-Based Polymers Analogous to Petroleum-Derived Ones for Packaging and Engineering Applications. *Polymers*, *12*(8), 1641. doi: 10.3390/polym12081641

Skibiński, S., Cichoń, E., Haraźna, K., Marcello, E., Roy, I., Witko, M., Ślósarczyk, A., Czechowska, J., Guzik, M., & Zima, A. (2021). Functionalized Tricalcium Phosphate and Poly(3-Hydroxyoctanoate) Derived Composite Scaffolds as Platforms for the Controlled Release of Diclofenac. *Ceramics International*, *47*(3), 3876–3883. doi: 10.1016/j.ceramint.2020.09.248

Solarz, D., Witko, T., Karcz, R., Malagurski, I., Ponjavic, M., Levic, S., Nesic, A., Guzik, M., Savic, S., & Nikodinovic-Runic, J. (2023). Biological and Physiochemical Studies of Electrospun Polylactid/Polyhydroxyoctanoate PLA/P(3HO) Scaffolds for Tissue Engineering Applications. *RSC Advances*, *13*(34), 24112–24128. doi: 10.1039/D3RA03021K

Suarez, A., Ford, E., Venditti, R., Kelley, S., Saloni, D., & Gonzalez, R. (2023). Is Sugarcane-Based Polyethylene a Good Alternative to Fight Climate Change? *Journal of Cleaner Production*, *395*, 136432. doi: 10.1016/J.JCLEPRO.2023.136432

Tharanathan, R. N. (2003). Biodegradable Films and Composite Coatings: Past, Present and Future. *Trends in Food Science and Technology*, *14*(3), 71–78. doi: 10.1016/S0924-2244(02)00280-7

Vandi, L. J., Chan, C. M., Werker, A., Richardson, D., Laycock, B., & Pratt, S. (2018). Wood-PHA Composites: Mapping Opportunities. *Polymers*, *10*(7), 1–15. doi: 10.3390/polym10070751

Velema, J., & Kaplan, D. (2006). Biopolymer-based Biomaterials as Scaffolds for Tissue Engineering. *Advances in Biochemical Engineering/Biotechnology*, *102*(July), 187–238. doi: 10.1007/10_013

Verlinden, R. A. J. A. J., Hill, D. J. J., Kenward, M. A. A., Williams, C. D. D., & Radecka, I. (2007). Bacterial Synthesis of Biodegradable Polyhydroxyalkanoates. *Journal of Applied Microbiology*, *102*(6), 1437–1449. doi: 10.1111/j.1365-2672.2007.03335.x

Vroman, I., & Tighzert, L. (2009). Biodegradable Polymers. *Materials*, *2*(2), 307–344. doi: 10.3390/ma2020307

Winter Is Coming: Plastic has to Go: A Case for Decreasing Plastic Production to Reduce the European Union's Dependence on Fossil Fuels and Russia (2022). *Ciel, September*. Retrieved from https://www.ciel.org/reports/winter-is-coming-plastic-has-to-go/

Witko, T. (2019). *Biophysical Characteristics and Cellular Studies of Polyhydroxyoctanoate (PHO) – Biodegradable and Biocompatible Polymer for Biomedical Applications*. Jagiellonian University, Krakow.

Witko, T., Guzik, M., Sofińska, K., Stepien, K., & Podobinska, K. (2018). Novel Biocompatible Polymers for Biomedical Applications. *Biophysical Journal*. doi: 10.1016/j.bpj.2017.11.2014

Witko, T., Solarz, D., Feliksiak, K., Haraźna, K., Rajfur, Z., & Guzik, M. (2020). Insights into in Vitro Wound Closure on Two Biopolyesters – Polylactide and Polyhydroxyoctanoate. *Materials, 13*(12), 2793. doi: 10.3390/ma13122793

Witko, T., Solarz, D., Feliksiak, K., Rajfur, Z., & Guzik, M. (2019). Cellular Architecture and Migration Behavior of Fibroblast Cells on Polyhydroxyoctanoate (PHO): A Natural Polymer of Bacterial Origin. *Biopolymers*. doi: 10.1002/bip.23324

Wu, C. S. (2012). Utilization of Peanut Husks as a Filler in Aliphatic–Aromatic Polyesters: Preparation, Characterization, and Biodegradability. *Polymer Degradation and Stability, 97*(11), 2388–2395. doi: 10.1016/J.POLYMDEGRADSTAB.2012.07.027

Yadav, H., Yadav, V., & Shah, G. (2015). Biomedical Biopolymers, their Origin and Evolution in Biomedical Sciences: A Systematic Review. *Journal of Clinical and Diagnostic Research, 9*(9), 21–25. doi: 10.7860/JCDR/2015/13907.6565

Yin, Q., Lu, C., Zhang, S., Liu, M., Du, K., Zhang, L., & Chang, G. (2019). Microporous Organic Hydroxyl-Functionalized Polybenzotriazole for Encouraging CO_2 Capture and Separation. *RSC Advances, 9*(39), 22604–22608. doi: 10.1039/C9RA03741A

Younes, B. (2017). Classification, Characterization, and the Production Processes of Biopolymers Used in the Textiles Industry. *The Journal of the Textile Institute, 108*(5), 674–682. doi: 10.1080/00405000.2016.1180731

13

THE LIFE CYCLE ASSESSMENT OF BIOPOLYMER AS AN EXAMPLE OF A SUSTAINABLE PRODUCT

Marcin Rychwalski and Tomasz Witko

13.1   Introduction

In the face of escalating environmental concerns and the urgent need for sustainable practices, industries worldwide are increasingly turning toward biopolymers as a promising solution (Patel & Blumberga, 2023). These biodegradable alternatives to conventional plastics offer the potential to mitigate the detrimental impact of traditional polymers on ecosystems and human health. However, the quest for sustainability demands a comprehensive understanding of the entire life cycle of these biopolymers – from raw material extraction to end-of-life disposal.

This chapter delves into the intricate world of biopolymer sustainability through the lens of life cycle assessment (LCA). By examining selected stages in a biopolymer's life cycle, we aim to shed light on its environmental footprint, resource consumption, and potential for improvement. As the urgency to make the transition toward a circular economy intensifies, it is becoming imperative to provide a critical evaluation of the sustainability claims associated with biopolymers (Dziuba et al., 2021). By conducting a thorough LCA, we can discern the true environmental implications of adopting biopolymers and pave the way for informed decision-making in pursuit of a greener future (Scanlon et al., 2013).

13.2   LCA analysis

A comprehensive Life Cycle Assessment (LCA) for biopolymers should encompass all stages of a product's life cycle, from raw material extraction up to end-of-life disposal, taking into consideration such factors as resource extraction,

DOI: 10.4324/9781032710693-14

manufacturing processes, distribution, product use, and disposal methods (Kim & Dale, 2005). Of paramount importance is an evaluation of the environmental impact of acquiring biomass feedstock, including land use, water consumption, energy inputs, and associated emissions during cultivation, thereby ensuring sustainable sourcing practices (Nitkiewicz et al., 2020). The biopolymer manufacturing process must be thoroughly examined, assessing energy consumption, emissions, waste generation, and the use of chemicals to identify opportunities for optimization and eco-friendly alternatives. LCA should also consider the transportation footprint, analyzing emissions caused by the transportation of raw materials to manufacture facilities as well as by the distribution of finished products to consumers, and thus covering the impact of the entire supply chain. End-of-life treatment options such as composting, recycling, and incineration should be evaluated to determine the environmental fate of biopolymers, taking into account such factors as biodegradability, compostability, and potential impacts on soil, water, and air quality.

13.3   Scope of study

Medium chain length-polyhydroxyalkanoates (mcl-PHAs) and polyhydroxybutyrate (PHB) are both biodegradable biopolymers synthesized by microorganisms from renewable feedstocks (Klinke et al., 1999; Verlinden et al., 2007). Both materials belong to the PHA group and represent a promising class of biodegradable polymers derived from renewable resources (Bugnicourt et al., 2014; Witko, 2019). However, their structural differences and production processes can lead to variations in their environmental impact, making a comparative LCA crucial for informed decision-making. An LCA was conducted for two biopolymers to evaluate their climate impact, taking into consideration greenhouse gas emissions. The analysis aimed to compare their environmental footprint throughout subsequent life cycle stages. Factors such as raw material production, manufacturing processes, use phase, and end-of-life disposal were taken into account. The study sheds light on the sustainability of these biopolymers and their potential role in mitigating climate change. The results of this analysis may provide decision-makers with valuable insights when choosing environmentally friendly materials.

The study compared two production scenarios for biopolymers, assessing their potential impact on global climate warming using the IPCC 2021 method. The comparison evaluated the proportion of PHA polyhydroxyalkanoate production from vegetable oil and poly(3-hydroxybutyrate) production from biodiesel. The study examined alternative production scenarios for these products in relation to environmental categories, which were published by Nitkiewicz et al. (2020). The aim of the study was to compare the global warming emission potential of both variants and to assess the possibility of CO_2 absorption as part of co-occurring processes. The comparison used a cradle-to-gate approach, ensuring that the same functional unit was used for both variants.

13.4 Microbial production of polyhydroxyalkanoates

Microorganisms, notably bacteria such as Pseudomonas, Bacillus, or Cupriavidus, are employed for PHA synthesis (Johnston et al., 2018; Mizuno et al., 2014). These microorganisms utilize the carbon source provided in the fermentation medium and convert it into PHA through enzymatic pathways. The fermentation process is carefully controlled, with such parameters as temperature, pH, oxygen supply, and nutrient availability to enhance PHA production yields. After fermentation, the microbial biomass containing PHA is harvested from the culture broth. Various methods such as solvent extraction, enzymatic digestion, or cell disruption techniques are employed to isolate PHA from cellular components (Favaro et al., 2019; Heinrich et al., 2012; Ibrahim et al., 2009). Solvent extraction using organic solvents such as chloroform or methylene chloride is commonly applied, followed by purification steps to obtain high-quality PHA. The extracted PHA undergoes purification processes to remove residual impurities and solvents, thus ensuring the desired polymer quality. Purification techniques may include precipitation, filtration, or chromatography. Once purified, PHA can be processed into various forms such as pellets, films, or fibers using conventional polymer processing techniques such as extrusion, injection molding, or compression molding and more advanced methods such as wet spinning, electrospinning, or additive techniques, for example 3D printing (Li et al., 2016; Malagurski et al., 2021; Solarz et al., 2023). Figure 13.1 shows a simplified diagram presenting the life cycle of biopolymers such as mcl-PHA and PHB, including the production, extraction, management, and utilization phases. The production yields of PHA vary depending on such factors as the chosen microbial

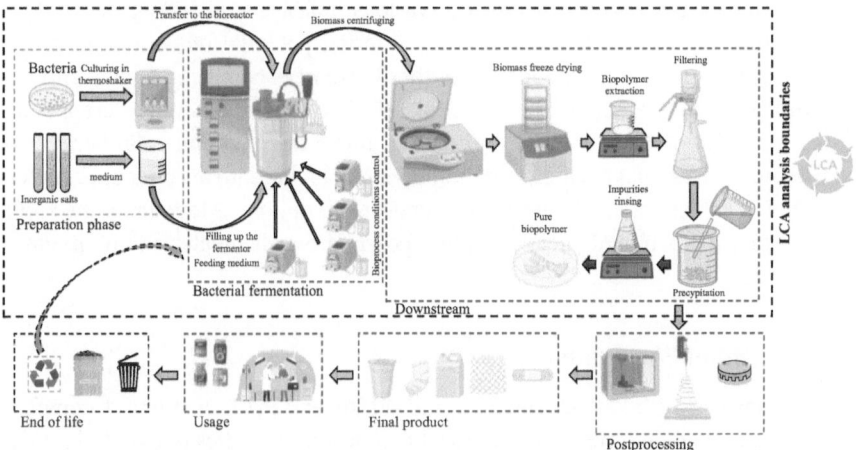

FIGURE 13.1   Schematic diagram of the life cycle of sample bacterial polymers, including the production, extraction, management, and utilization phases. The boundaries of the LCA analysis are marked.

Source: Author's own diagram.

strain, the fermentation conditions, and the substrate used. However, significant advances in bioprocess engineering and strain optimization have led to remarkable yields in recent years. For instance, production yields of up to 70% of cell dry weight have been achieved for mcl-PHA (Kacanski et al., 2023; Reddy et al., 2022; Silva et al., 2021; Zhu et al., 2022), while PHB yields exceeding 80% have been reported in certain studies (Ibrahim et al., 2009; Trakunjae et al., 2021).

13.4.1 Raw material acquisition

Polyhydroxyalkanoates are typically produced from renewable resources such as fatty acids derived from vegetable oils or waste streams, while PHB is synthesized from glycerol or sugars obtained from crops such as corn or sugarcane (Hankermeyer & Tjeerdema, 1999). LCA analysis must consider such factors as land use, water consumption, and energy inputs associated with feedstock cultivation, extraction, and processing.

13.4.2 Manufacturing process

The production of both mcl-PHAs and PHB involves the microbial fermentation of carbon substrates, followed by downstream processing for polymer recovery. LCA should assess the energy requirements, chemical usage, and waste generation during these manufacturing processes, taking into account variations in resource consumption and emissions (Nitkiewicz et al., 2020).

13.4.3 Product applications

Both materials can find applications in various industries, including the packaging, biomedical, and agricultural sectors, offering potential reductions in greenhouse gas emissions and dependence on fossil resources compared to conventional plastics (Baranwal et al., 2022; Cruz et al., 2022; Gupta et al., 2022; Perera et al., 2023; Rababah et al., 2020). LCA should evaluate the environmental benefits of substituting traditional plastics with biopolymers, taking into account such factors as material performance, durability, and end-of-life options. More about the applications of various biopolymers including polyhydroxyalkanoates can be found in Chapter 12.

13.4.4 End-of-life scenario

Biodegradability is a key advantage of mcl-PHAs and PHB, enabling decomposition in natural environments or industrial composting facilities (Cruz et al., 2022; Kuruppalil, 2011). LCA should assess the fate of biopolymers after use, taking into account such factors as degradation rates, by-products, and potential impacts on soil quality and marine ecosystems.

13.5 LCA analysis

The purpose of this study is to determine the potential environmental impact of the produced polymers, by comparing the greenhouse potential of the raw materials and materials used in their production. The study aims to identify key aspects that can improve the environmental footprint of the process and enable conscious choices regarding raw materials and processes used in their production (ISO 14040:2006(en), *Environmental Management, Life Cycle Assessment, Principles and Framework*, 2006).

The study is limited to the manufacturing processes and market for raw materials, chemical components, and other inputs required for obtaining the polymer. It does not include the production of manufacturing equipment, material postprocessing, usage, or disposal scenarios (Figure 13.1). The study will only compare two product variants, covering the basic inputs and outputs of the inventory phase. The analysis is followed with the production of a specific amount of product, taking into account the energy and water required in the production and downstream phase. The packaging of ingredients required for production, chemicals, and organizational and office production facilities are also not considered in this study.

This chapter aims at individuals within the scientific community involved in production processes and their organization. Its findings will facilitate the identification of environmental impacts and inform future decisions regarding raw materials and production processes.

13.5.1 Methodology

The study was conducted using the IPCC 2021 (Intergovernmental Panel on Climate Change, 2023) methodology for assessing environmental damage. This methodology utilizes the global warming potential (GWP) indicator, which also constituted a part of the earlier EPD 2018 (*EN ISO 14025:2010*, 2010) method developed by the IPCC to evaluate the impact of various greenhouse gases on global warming. The assessment covers a period of 100 years, during which GWP values are calculated based on the greenhouse potential of the gas compared to carbon dioxide (CO_2).

The IPCC 2021 method provides various characterization factors, which result in six method variants that quantify GWP and two methods that quantify global temperature potential (GTP). The assessment always uses versions of the same method that take into account carbon dioxide with an update of its absorption possibility over a 100-year period. For GWP, different time horizons are available: 20 years, 100 years (default), and 500 years. It is important to note that GWP100 is recommended as the default metric for the Global Guidelines on Life Cycle Impact Assessment Metrics and Methods (Frischknecht et al., 2016) and GWP20, while GTP100 is recommended for sensitivity analysis.

The analysis takes into account IPCC 2021 GWP100 (including CO_2 absorption), that is, the GWP IPCC climate change factors over a 100-year time horizon. This method explicitly includes carbon dioxide emissions and regards carbon dioxide absorption as biogenic.

The IPCC has modeled the factors that contribute to global warming and temperature change potential, along with the climate carbon feedback. However, the formation of nitrous oxide caused by nitrogen emissions is not included in the model. Additionally, the radiative forcing caused by short-term climatic factors such as nitrogen oxides, carbon monoxide (CO), volatile organic compounds, soot, organic carbon, and sulfur oxides has not been considered in accordance with UNEP (Niblick & Initiative, 2019) recommendations. The indirect effects of CO emissions are not taken into account in the results obtained (Forster et al., 2023). Finally, the impact assessment results combine all the impacts and categories together.

A simplified LCA of polymer production was conducted using the IPCC 2021 method with uptake. The results for GWP were presented in the form of CO_2 equivalent produced, taking into account CO_2 absorption in biological processes (Nitkiewicz et al., 2020). The method and scope of impacts included in the impact assessment are discussed more broadly in Chapter 4. This chapter presents the role of LCA in the design of sustainable products and describes the most important, updated environmental assessment methods employed. The assessment was performed separately for two types of polymers and also in comparison with the same production volume. The study was conducted on the basis of SimaPro software and the latest updated versions of the environmental databases Ecoinvent 3.10 (Ruiz, 2023), Agri-footprint (Nitkiewicz et al., 2020), Agrybalyse (Auberger et al., 2022), and Industry Data 2.0 (Nitkiewicz et al., 2020).

The functional unit was assumed to be the production of 1 kg of both tested microbiological polymers as part of small-batch production.

13.6   Results

In the study, the functional unit was considered to be the amount of polymer produced at a time using a 200-liter fermenter. The average amount of polymer produced using a 200-liter fermenter was 0.78 kg in the case of mcl-PHA and 1.44 kg in the case of PHB. The production process for both variants was divided into five main elements, with the contributions of each element to the environmental impact being specified. These elements include the preparation of the medium, fermentation, trace elements, water for sterilization, and electricity. Data for this study was obtained from direct energy and water consumption measurements for fermentation processes performed using a Sartorius Biostat D DCU fermenter. All auxiliary equipment consumption was also directly measured during the running process.

According to mcl-PHA, the primary cause of environmental impact in terms of carbon dioxide equivalent is the energy used in production (Figure 13.2). This energy is derived from the energy mix available on the energy products market

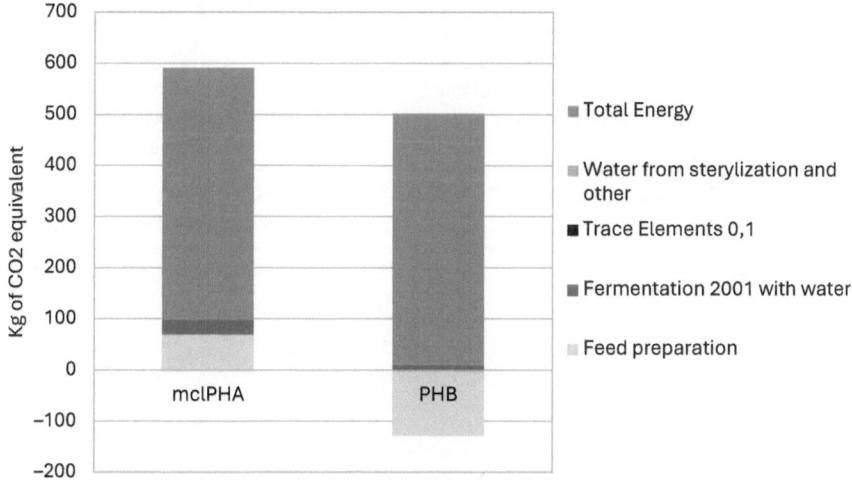

FIGURE 13.2 Coal footprint compare of single production of mcl-PHA and PHB.

Source: Author's own research.

in Poland, which is closely associated with the place of production. The energy mix in Poland in 2024 will mainly consist of hard coal (43.26%) and brown coal (19.77%), with wind farms contributing 19.55% and other renewables another 3.96%. The composition of electricity production in February 2024 further confirms this mix. The scale of these impacts is directly related to the energy mix in a given region. However, the amount of energy consumed also plays a significant role and, in both cases, the impact is approximately 492 kg of CO_2 with an assumed average consumption of 515 kWh (Table 13.1).

The comparison being made concerns the production of polymers from a 200-liter fermenter. mcl-PHA yields 0.78 kg of product, while PHB yields 1.44 kg. However, in terms of overall GWP, the result for mcl-PHA is higher than for PHB, that is, 555.73 kg eq CO_2 compared to PHB's 373.74 kg eq CO_2. The medium preparation process is the second crucial phase after energy consumption.

The medium used in the production process has a significant impact on the environment, but it can also absorb carbon dioxide with CO_2 uptake (Figure 13.3). A Sankey chart (Figure 13.4) can visualize the individual component processes, with fatty acids being the leading contributor (up to 5%). The use of plant-based fatty acids derived from palm oil and soybean oil is taken into account. While palm oil has the potential to absorb carbon dioxide, which offsets part of the impact (-4.98 kg eq CO_2), the impact of soybean production is greater than the absorption.

Soybean production processes emit 72.3 kg CO_2 equivalent, while their absorption potential is only 7.2 kg CO_2 equivalent (Figure 13.4). This results in a global warming impact of 65.1 kg CO_2 equivalent. The main impact in the production process is generated by the energy used, in the form of air emissions, the use of

TABLE 13.1 Impact contributions for production stages (kg of CO_2 equivalent)

kg CO_2-eq	Total	Feed and medium preparation	Fermentation volume 200 L	Trace elements	1. Water for sterilization and other processes	Total energy
mcl-PHA	590.41908	70.323818	28.189619	0.071562082	0.059516431	491.77457
PHB	373.74519	−128.06193	9.9014724	0.071562082	0.059516431	491.77457

Source: Author's own research.

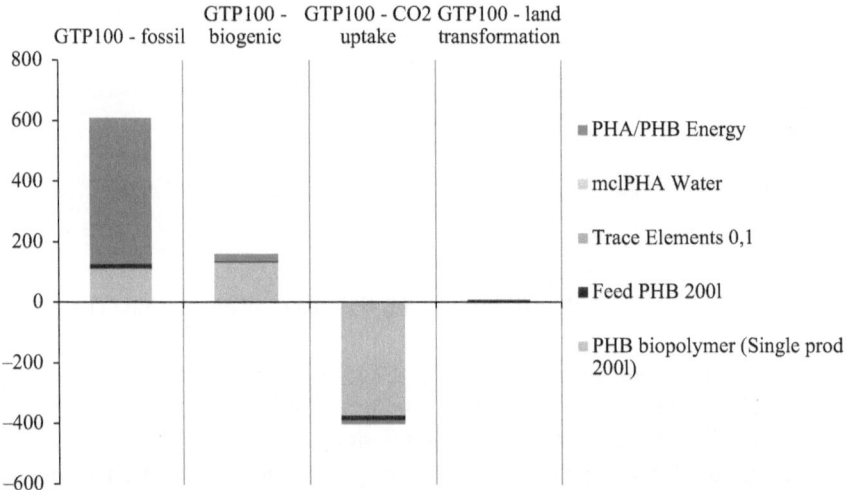

FIGURE 13.3   Shares of process groups in the environmental result for mcl-PHA production.

Source: Author's own research.

primary raw materials, and the fuels consumed for this purpose. Soybean production has a huge impact through its exploitation of large areas of land.

To reduce the greenhouse potential of manufactured products, it has been suggested that manufacturers look for raw materials and ingredients with greater absorption potential. This will help them balance the effect of greenhouse gases. The production of PHB, the second polymer, has a lower CO_2 equivalent result. This is due to the even greater absorption potential of the nutrients used.

The combined GWP greenhouse potential (608.64 kg CO_2-eq) of fossil material demand for PHB in terms of energy consumed and production processes is higher than for mcl-PHA (555.73 kg CO_2-eq) (Figure 13.3 vs Figure 13.5). The biogenic emission rate is also higher (160.04 kg CO_2-eq vs 92.75 kg CO_2-eq) (Table 13.2). However, the greenhouse emission factor in the case of land use is much lower, which, when combined with the CO_2 absorption potential from ingredient

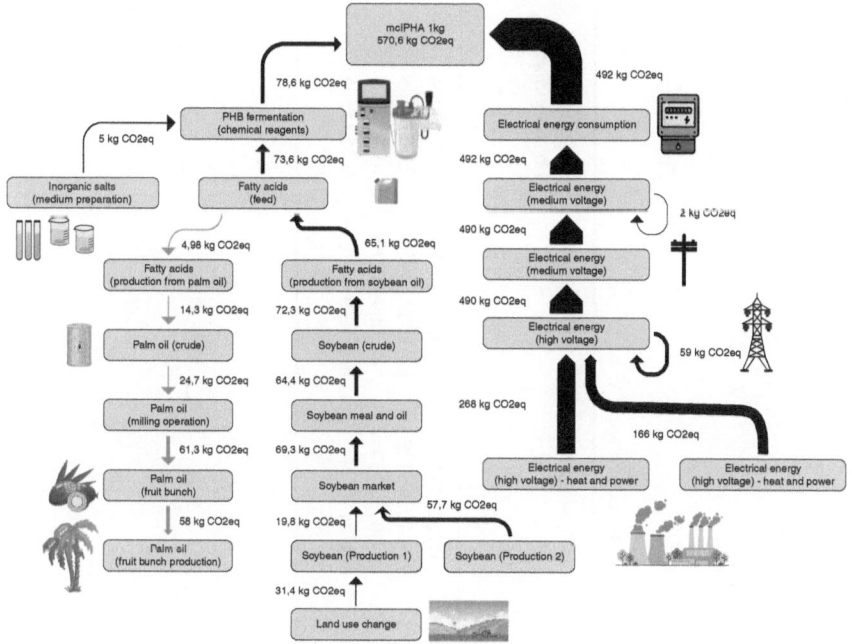

FIGURE 13.4   Sankey plot for the mcl-PHA production process.

Source: Author's own study.

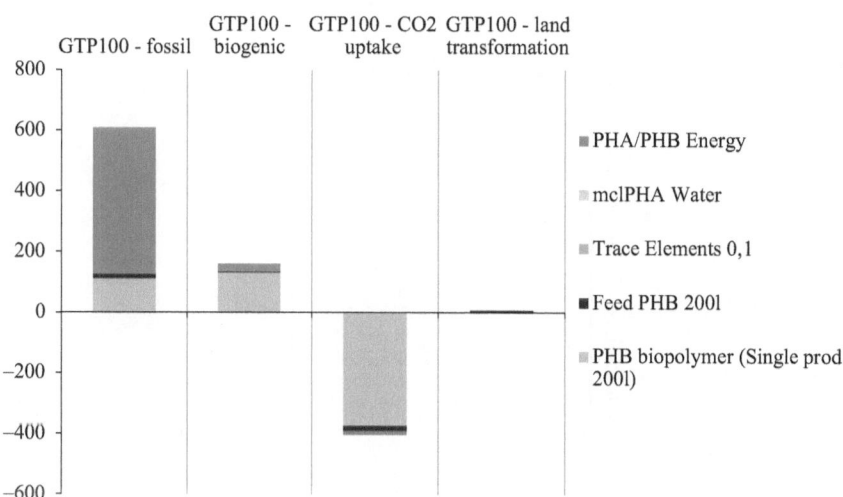

FIGURE 13.5   Shares of different process groups in the environmental result for PHB production.

Source: Author's own research.

TABLE 13.2 Impact contributions of different production stages in GWP categories (kg of CO_2 equivalent)

Impact category	Unit	Total	2. PHB biopolymer (single prod 200 L)	Feed PHB 200 L	Trace elements 0,1	mcl-PHA water	Energy consumption
GTP100-fossil	kg CO_2-eq	608.64129	112.94764	14.29189	0.071032613	0.040567696	481.29016
GTP100-biogenic	kg CO_2-eq	160.03879	132.51696	3.9002186	0.0040249749	0.019891667	23.597689
GTP100-CO_2 uptake	kg CO_2-eq	−402.41108	−373.88639	−15.254259	−0.0035813754	−0.00099477796	−13.265854
GTP100-land transformation	kg CO_2-eq	7.4761907	0.35985538	6.9636231	8.59E−05	5.18E−05	0.15257458

Source: Author's own research.

production, gives a more favorable result (373.74 kg CO_2-eq vs. 590.42 kg CO_2-eq) (Table 13.1).

The Sankey diagram shows the high energy utilization associated with PHB fermentation processes and its high share in the final result. The energy processes visible in the diagram are used in the production process and are also part of the system for generating and transporting electricity from the power source.

In PHB microbial production glycerol, produced from rapeseed oil (Figure 13.6), is used as a feeding medium. The absorption properties of rapeseed plants allow for a significant reduction in the greenhouse potential of PHB polymer production. Without this potential, the GWP results would be higher for PHB than for mcl-PHA. The feeding farm process also involves the use of glycerol, which further reduces the impact of CO_2 eq. This is shown in Table 13.2. The carbon footprint is more favorable due to greater production efficiency when the same 200-liter fermenter capacity is used.

13.6.1 Comparing the same quantity of products

To compare the production values for the same quantity of product, 10 kg of polymer was used. The functional unit was changed in the next study to enable a direct comparison of the products.

In this way, we can confirm the proportion of production results and impacts when the same product quantity is used. Differences in production efficiency should become more visible and comparisons between products more transparent. We can see the proportions differ in terms of GWP, including in terms of less land use, greater absorption potential, as well as a greater need for primary sources for both polymers (Figure 13.7).

In a previous comparison made between PHB polymer and mcl-PHA, it was found that PHB is more efficient with the same resources. Furthermore, PHB has the potential to absorb carbon dioxide. Although mcl-PHA has a higher absorption potential per unit weight, the overall CO_2 equivalent is lower for PHB due to its more efficient production system (Table 13.3). This highlights the importance of improving production efficiency to maximize output while minimizing inputs.

Table 13.4 shows that the advantage of higher production efficiency is even greater when translated into the carbon footprint index. The absorption potential of production and a lower land transformation indicator also contribute to the end result, in which PHB production gains a significant advantage (Table 13.3).

13.7 Key findings and challenges for biopolymers in the future

Biopolymers represent a promising alternative to traditional fossil fuel-based plastics, offering a pathway toward mitigating carbon emissions and reducing energy consumption in various industries. Unlike conventional plastics originating from non-renewable resources, biopolymers are derived from biomass sources, such as plants, animals, or microorganisms and present a more sustainable option

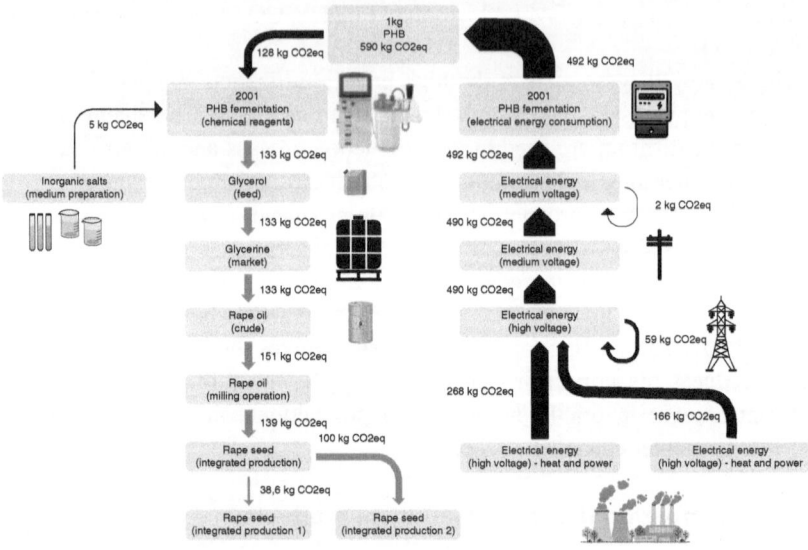

FIGURE 13.6 Sankey plot for mcl-PHB production.

Source: Author's own study.

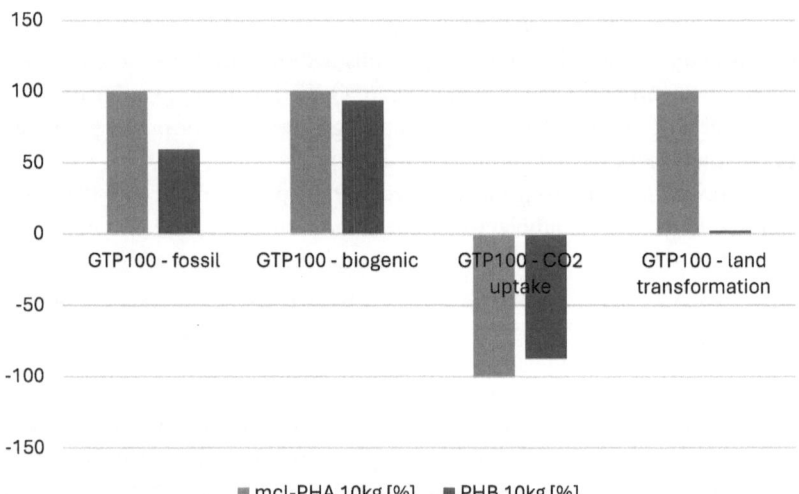

FIGURE 13.7 A comparison of 10 kg of products in different GWP categories.

Source: Author's own research using SimaPro software.

TABLE 13.3 Total coal footprint of 10 kg polymers production

Damage category	Unit	mcl-PHA 10 kg	PHB 10 kg
GTP100 incl. CO_2 biogenic	kg CO_2-eq	7569.1727	2597.5291

Source: Author's own research.

TABLE 13.4 Coal footprint of 10 kg polymer production in different damage categories

Impact category	Unit	mcl-PHA 10 kg	PHB 10 kg
GTP100-fossil	kg CO_2-eq	7124.4532	4230.057
GTP100-biogenic	kg CO_2-eq	1189.0689	1112.2696
GTP100-CO_2 uptake	kg CO_2-eq	−3204.3087	−2796.757
GTP100-land transformation	kg CO_2-eq	2459.9593	51.959526

Source: Author's own research.

for material production (Cruz et al., 2022; Gupta et al., 2022). One of the key advantages of biopolymers lies in their significantly lower carbon emissions throughout the production process. The carbon footprint of biopolymers is inherently lower due to their renewable origin, as biomass sources absorb CO_2 during growth, offsetting emissions associated with their cultivation and processing. This contrasts sharply with fossil fuel–based plastics, which not only emit CO_2 during extraction and processing but also contribute to overall greenhouse gas emissions through combustion and degradation (Blunt et al., 2018). Moreover, biopolymer production typically requires less energy compared to conventional plastics. The energy intensity of biopolymer manufacturing processes is often reduced due to the lower processing temperatures required, as well as the use of renewable energy sources in some cases. Additionally, biopolymers generally require fewer chemical inputs and less complex processing techniques, further contributing to energy efficiency and reducing the environmental impact. The environmental benefits of biopolymers extend beyond carbon emissions and energy consumption. Their renewable nature reduces dependency on finite fossil fuel resources, thereby mitigating concerns related to resource depletion and as a consequence enhancing long-term sustainability (Hottle et al., 2017). Additionally, biopolymers offer the potential to alleviate plastic pollution and its associated environmental hazards by providing biodegradable and compostable alternatives, thus reducing waste accumulation in landfills and oceans.

However, it is important to be aware of the fact that the environmental performance of biopolymers can vary depending on such factors as feedstock sourcing, cultivation practices, manufacturing methods, and end-of-life treatment options. Careful consideration and evaluation with tools such as LCA are crucial for accurately assessing the overall sustainability of biopolymer products and identifying areas for improvement.

The competition with food crops for land is a critical concern when it comes to the sustainable production of biopolymers (Kuruppalil, 2011; Mtibe et al., 2021; Sherwood et al., 2017). As the demand for bio-based materials increases, there is a risk of encroaching upon agricultural land traditionally reserved for food cultivation. This competition raises ethical questions regarding food security, land use practices, and equitable resource distribution. Striking a balance between the cultivation of biomass for biopolymers and food production is essential to offset any potential conflicts and ensure global food security. Furthermore, the limited scalability of production inhibits the widespread availability and affordability of biopolymers (Kuruppalil, 2011; Mtibe et al., 2021; Sherwood et al., 2017). Unlike conventional plastics, which benefit from mature and highly efficient manufacturing processes, biopolymer production often relies on nascent technologies with lower economies of scale. Scaling up production to meet growing demand while maintaining environmental sustainability presents technical and logistical challenges. Research and investment in innovative production methods are necessary to enhance efficiency, reduce costs, and increase the viability of large-scale biopolymer production. The lack of infrastructure for collection and recycling poses significant obstacles to the circularity of biopolymer materials. Existing recycling systems primarily cater to fossil fuel–based plastics, rendering them incompatible with biodegradable or compostable biopolymers (Kuruppalil, 2011; Mtibe et al., 2021; Sherwood et al., 2017). Without adequate facilities and processes for the separate collection, sorting, and processing of biopolymer waste, these materials risk ending up in landfills or contaminating recycling streams, negating their environmental benefits. Developing a robust infrastructure for biopolymer waste management, including composting facilities and specialized recycling technologies, is crucial for closing the loop and realizing the full potential of biopolymers as sustainable alternatives.

The end-of-life behavior of biopolymers represents a critical aspect when assessing their overall sustainability. Unlike traditional plastics derived from fossil fuels, the fate of biopolymers at the end of their useful life can vary significantly based on several factors, including composition, processing methods, and environmental conditions (Kuruppalil, 2011; Mtibe et al., 2021; Sherwood et al., 2017). This variability necessitates careful consideration of waste management strategies to ensure that their environmental benefits are fully realized. The composition of biopolymers plays a pivotal role in determining their end-of-life behavior. Biopolymers can be categorized into different types, each with distinct properties and degradation characteristics. For instance, while some biopolymers are designed to be compostable under specific conditions, others may require industrial facilities for proper decomposition. Understanding these differences is crucial for implementing appropriate waste management practices. The processing methods used in manufacturing biopolymers can influence their biodegradability and recyclability. Factors such as temperature, pressure, and additives utilized during processing can affect the molecular structure of biopolymers, thereby impacting their ability to degrade or be recycled efficiently. Therefore, selecting environmentally friendly processing techniques and minimizing the use of harmful additives are essential factors that need to be taken into account when endeavoring to enhance the end-of-life performance of biopolymers. Environmental conditions, including temperature, humidity, and microbial activity, play a significant role in the degradation of biopolymers. For instance, compostable biopolymers may require specific conditions, such as elevated temperatures and adequate moisture levels, to facilitate microbial decomposition effectively. In contrast, biopolymers intended for recycling may need to be shielded from environmental factors that could compromise their structural integrity. Thus, tailoring waste management strategies to suit prevailing environmental conditions is essential for optimizing the end-of-life outcomes of biopolymers.

Given the diverse range of biopolymers and their unique end-of-life behaviors, a one-size-fits-all approach to waste management is impractical. Instead, a holistic and context-specific approach is required, taking into account the specific properties of biopolymers, local infrastructure, and environmental considerations. Collaboration between stakeholders, including policymakers, manufacturers, waste management facilities, and consumers, is essential to develop and implement effective waste management strategies that maximize the environmental benefits of biopolymers while minimizing their ecological footprint.

13.8   Conclusions

LCA analysis of mcl-PHA and PHB provides valuable insights into the CO_2 emissions and carbon footprint associated with these biopolymers, aiding informed decision-making aimed at developing sustainable materials. By identifying hotspots and potential mitigation strategies across the life cycle, LCA contributes to the advancement of environmentally friendly biopolymer production processes, ultimately supporting the transition toward a circular and low-carbon economy.

The analysis focused on the preparation processes, microbial production, and downstream processes of two biopolymers. In these stages, it is evident that the factor that has the greatest impact on their carbon footprint is energy consumption. The devices used in this process consume a considerable amount of electrical energy due to the need to establish and maintain optimal conditions for bacterial growth and polymer accumulation throughout the process duration.

In order to compare two product variants, further analysis of the whole life cycle is required. The research results for these products indicate that they can help balance the greenhouse effect, as they focus on energy sources, production efficiency, and ingredients of plant origin that can absorb carbon.

Continued research and efforts at improvement in both biopolymer production as well as recycling methodologies and analysis will further enhance our understanding of the environmental implications of bioplastics.

LCA of biopolymers holds enormous promise as a tool for shaping future sustainability practices across various industries. By meticulously analyzing the environmental impacts of biopolymer production, usage, and disposal, LCA provides invaluable insights into how to reduce carbon footprints and promote eco-friendly practices. Future applications of LCA for biopolymers will encompass a wide array of sectors including packaging, textiles, automotive, and construction. As society gravitates toward more sustainable alternatives, LCA will serve as an aid in decision-making processes, guiding companies toward selecting the most environmentally friendly biopolymer options. Furthermore, LCA can shed light on innovations in biopolymer manufacturing processes, leading to the development of more efficient production methods with reduced energy consumption and waste generation. This could potentially revolutionize the biopolymer industry, making it even more competitive against traditional petrochemical-based polymers. The implications of the widespread adoption of LCA for biopolymers are profound. It could trigger a paradigm shift toward a circular economy, in which biopolymers are not only derived from renewable resources but also recycled and reused efficiently, thus minimizing resource depletion and waste accumulation. Ultimately, the integration of LCA at every stage of biopolymer production and utilization is pivotal for shaping the future trajectory of this industry. Through informed decision-making and conscientious consumption, LCA has the potential to bring about a more sustainable and resilient future for both industry and the environment.

Acknowledgement

The publication/article presents the result of the Project no 070/ZJE/2024/POT financed from the subsidy granted to the Krakow University of Economics.

References

Auberger, J., Ayari, N., Ceccaldi, M., Cornelus, M., & Geneste, C. (2022). Agribalyse Change Report 3.0/3.1, Written by INRAE teams–MEANS Platform, EVEA. ADEME 2022 Edition, https://3613321239-files.gitbook.io/~/files/v0/b/gitbook-x-prod.appspot.com/o/spaces%2F-M7H-JTDnDsswmNDPy-z%2Fuploads%2FI5Kr1AJBSxaoEUMtnRdT%2FChange Report_oct2022.pdf?alt=media&token=654009df-db4e-4d0c-874d-97c0f76a9741

Baranwal, J., Barse, B., Fais, A., Delogu, G. L., & Kumar, A. (2022). Biopolymer: A Sustainable Material for Food and Medical Applications. *Polymers*, *14*(5), 983. doi: 10.3390/polym14050983

Blunt, W., Levin, D. B., & Cicek, N. (2018). Bioreactor Operating Strategies for Improved Polyhydroxyalkanoate (PHA) Productivity. *Polymers*, *10*(11). doi: 10.3390/polym10111197

Bugnicourt, E., Cinelli, P., Lazzeri, A., & Alvarez, V. (2014). Polyhydroxyalkanoate (PHA): Review of Synthesis, Characteristics, Processing and Potential Applications in Packaging. *Express Polymer Letters*, *8*(11), 791–808. doi: 10.3144/expresspolymlett.2014.82

Cruz, R. M. S., Krauter, V., Krauter, S., Agriopoulou, S., Weinrich, R., Herbes, C., Scholten, P. B. V., Uysal-Unalan, I., Sogut, E., Kopacic, S., Lahti, J., Rutkaite, R., & Varzakas, T. (2022). Bioplastics for Food Packaging: Environmental Impact, Trends and Regulatory Aspects. *Foods*, *11*(19), 3087. doi: 10.3390/foods11193087

Dziuba, R., Kucharska, M., Madej-Kiełbik, L., Sulak, K., & Wiśniewska-Wrona, M. (2021). Biopolymers and Biomaterials for Special Applications within the Context of the Circular Economy. *Materials*, *14*(24), 7704. doi: 10.3390/ma14247704

Favaro, L., Basaglia, M., & Casella, S. (2019). Improving Polyhydroxyalkanoate Production from Inexpensive Carbon Sources by Genetic Approaches: A Review. *Biofuels, Bioproducts and Biorefining*, *13*(1), 208–227. doi: 10.1002/bbb.1944

Forster, P., Storelvmo, T., Armour, K., Collins, W., Dufresne, J. L., Frame, D., Lunt, D. J., Mauritsen, T., M. D. Palmer, M. D. Watanabe, M., Wild, M., & Zhang, H. (2023). The Earth's Energy Budget, Climate Feedbacks and Climate Sensitivity. In Intergovernmental Panel on Climate Change (Ed.), *Climate Change 2021 – The Physical Science Basis: Working Group I Contribution to the Sixth Assessment Report of the Intergovernmental Panel on Climate Change* (pp. 923–1054). Cambridge: Cambridge University Press. doi: 10.1017/9781009157896.009

Frischknecht, R., Fantke, P., Tschümperlin, L., Niero, M., Antón, A., Bare, J., Boulay, A.-M., Cherubini, F., Hauschild, M. Z., Henderson, A., Levasseur, A., McKone, T. E., Michelsen, O., i Canals, L. M., Pfister, S., Ridoutt, B., Rosenbaum, R. K., Verones, F., Vigon, B., & Jolliet, O. (2016). Global Guidance on Environmental Life Cycle Impact Assessment Indicators: Progress and Case Study. *The International Journal of Life Cycle Assessment, 21*(3), 429–442. doi: 10.1007/s11367-015-1025-1

Gupta, I., Cherwoo, L., Bhatia, R., & Setia, H. (2022). Biopolymers: Implications and Application in the Food Industry. *Biocatalysis and Agricultural Biotechnology, 46*, 102534. doi: 10.1016/j.bcab.2022.102534

Hankermeyer, C. R., & Tjeerdema, R. S. (1999). Polyhydroxybutyrate: Plastic Made and Degraded by Microorganisms. *Reviews of Environmental Contamination and Toxicology, 159*, 1–24. doi: 10.1007/978-1-4612-1496-0_1

Heinrich, D., Madkour, M. H., Al-Ghamdi, M. A., Shabbaj, I. I., & Steinbüchel, A. (2012). Large Scale Extraction of Poly(3-Hydroxybutyrate) from Ralstonia Eutropha H16 Using Sodium Hypochlorite. *AMB Express, 2*(1), 1–6. doi: 10.1186/2191-0855-2-59

Hottle, T. A., Bilec, M. M., & Landis, A. E. (2017). Biopolymer Production and End of Life Comparisons Using Life Cycle Assessment. *Resources, Conservation and Recycling, 122*, 295–306. doi: 10.1016/j.resconrec.2017.03.002

Ibrahim, M. H. A. A., Steinbüchel, A., & Steinbüchel, A. (2009). Poly(3-Hydroxybutyrate) Production from Glycerol by Zobellella Denitrificans MW1 Via High-Cell-Density Fed-Batch Fermentation and Simplified Solvent Extraction. *Applied and Environmental Microbiology, 75*(19), 6222–6231. doi: 10.1128/AEM.01162-09

Intergovernmental Panel on Climate Change (2023). *Climate Change 2021 – The Physical Science Basis*. Cambridge University Press. doi: 10.1017/9781009157896

International Organization for Standardization. *ISO 14040: Environmental Management – Life Cycle Assessment – Principles and Framework*. 2006. ISO. https://www.iso.org/standard/37456.html

ISO 14025:2010. *Environmental Labels and Declarations – Type III Environmental Declarations – Principles and Procedures*. Geneva: International Organization for Standardization, 2010.

Johnston, B., Radecka, I., Hill, D., Chiellini, E., Ilieva, V. I., Sikorska, W., Musioł, M., Zięba, M., Marek, A. A., Keddie, D., Mendrek, B., Darbar, S., Adamus, G., & Kowalczuk, M. (2018). The Microbial Production of Polyhydroxyalkanoates from Waste Polystyrene Fragments Attained Using Oxidative Degradation. *Polymers, 10*(9). doi: 10.3390/polym10090957

Kacanski, M., Stelzer, F., Walsh, M., Kenny, S., O'Connor, K., & Neureiter, M. (2023). Pilot-Scale Production of mcl-PHA by Pseudomonas Citronellolis Using Acetic Acid as the Sole Carbon Source. *New Biotechnology, 78*, 68–75. doi: 10.1016/j.nbt.2023.10.003

Kim, S., & Dale, B. E. (2005). Life Cycle Assessment Study of Biopolymers (Polyhydroxyalkanoates) Derived from No-Tilled Corn. *International Journal of Life Cycle Assessment, 10*(3), 200–210. doi: 10.1065/lca2004.08.171

Klinke, S., Ren, Q., Witholt, B., & Kessler, B. (1999). Production of Medium-Chain-Length Poly(3-Hydroxyalkanoates) from Gluconate by Recombinant Escherichia Coli. *Applied and Environmental Microbiology*, *65*(2), 540–548. doi: 10.3181/00379727-20-248

Kuruppalil, Z. (2011). Green Plastics: An Emerging Alternative for Petroleum-Based Plastics. *International Journal of Engineering Research & Innovation*, *3*(1), 59–64. doi: 978-1-60643-379-9

Li, Z., Yang, J., & Loh, X. J. (2016). Polyhydroxyalkanoates: Opening Doors for a Sustainable Future. *NPG Asia Materials*, *8*(4), e265–20. doi: 10.1038/am.2016.48

Malagurski, I., Frison, R., Maurya, A. K., Neels, A., Andjelkovic, B., Senthamaraikannan, R., Babu, R. P., O'Connor, K. E., Witko, T., Solarz, D., & Nikodinovic-Runic, J. (2021). Polyhydroxyoctanoate Films Reinforced with Titanium Dioxide Microfibers for Biomedical Application. *Materials Letters*, *285*, 129100. doi: 10.1016/j.matlet.2020.129100

Mizuno, S., Katsumata, S., Hiroe, A., & Tsuge, T. (2014). Biosynthesis and Thermal Characterization of Polyhydroxyalkanoates Bearing Phenyl and Phenylalkyl Side Groups. *Polymer Degradation and Stability*, *109*, 379–384. doi: 10.1016/j.polymdegradstab.2014.05.020

Mtibe, A., Motloung, M. P., Bandyopadhyay, J., & Ray, S. S. (2021). Synthetic Biopolymers and Their Composites: Advantages and Limitations—An Overview. *Macromolecular Rapid Communications*, *42*(15). doi: 10.1002/marc.202100130

Niblick, B., & United Nations Environment Programme (UNEP). (2019). *Global Guidance for Life Cycle Impact Assessment Indicators (GLAM): Volume 2*. UNEP.

Nitkiewicz, T., Wojnarowska, M., Sołtysik, M., Kaczmarski, A., Witko, T., Ingrao, C., & Guzik, M. (2020). How Sustainable Are Biopolymers? Findings from a Life Cycle Assessment of Polyhydroxyalkanoate Production from Rapeseed-Oil Derivatives. *Science of the Total Environment*, *749*, 141279. doi: 10.1016/j.scitotenv.2020.141279

Patel, N., & Blumberga, D. (2023). Insights of Bioeconomy: Biopolymer Evaluation Based on Sustainability Criteria. *Environmental and Climate Technologies*, *27*(1), 323–338. doi: 10.2478/rtuect-2023–0025

Perera, K. Y., Jaiswal, A. K., & Jaiswal, S. (2023). Biopolymer-Based Sustainable Food Packaging Materials: Challenges, Solutions, and Applications. *Foods*, *12*(12), 2422. doi: 10.3390/foods12122422

Rababah, M. M., & AL- Oqla, F. M. (2020). Biopolymer Composites and Sustainability. In *Advanced Processing, Properties, and Applications of Starch and Other Bio-Based Polymers* (pp. 1–10). Elsevier. doi: 10.1016/B978-0-12-819661-8.00001-9

Reddy, V. U. N., Ramanaiah, S. V., Reddy, M. V., & Chang, Y.-C. (2022). Review of the Developments of Bacterial Medium-Chain-Length Polyhydroxyalkanoates (mcl-PHAs). *Bioengineering*, *9*(5), 225. doi: 10.3390/bioengineering9050225

Ruiz, M. E. (2023). *Documentation of Changes Implemented in the Ecoinvent Database v3.10 (2023.12.14)*.

Scanlon, K. A., Gray, G. M., Francis, R. A., Lloyd, S. M., & LaPuma, P. (2013). The Work Environment Disability-Adjusted Life Year for Use with Life Cycle Assessment: A Methodological Approach. *Environmental Health*, *12*(1), 21. doi: 10.1186/1476-069X-12–21

Sherwood, J., Clark, J. H., Farmer, T. J., Herrero-Davila, L., & Moity, L. (2017). Recirculation: A New Concept to Drive Innovation in Sustainable Product Design for Bio-Based Products. *Molecules*, *22*(1). doi: 10.3390/molecules22010048

Silva, J. B., Pereira, J. R., Marreiros, B. C., Reis, M. A. M., & Freitas, F. (2021). Microbial Production of Medium-Chain Length Polyhydroxyalkanoates. *Process Biochemistry*, *102*, 393–407. doi: 10.1016/j.procbio.2021.01.020

Solarz, D., Witko, T., Karcz, R., Malagurski, I., Ponjavic, M., Levic, S., Nesic, A., Guzik, M., Savic, S., & Nikodinovic-Runic, J. (2023). Biological and Physiochemical Studies of Electrospun Polylactid/Polyhydroxyoctanoate PLA/P(3HO) Scaffolds for Tissue Engineering Applications. *RSC Advances, 13*(34), 24112–24128. doi: 10.1039/D3RA03021K

Trakunjae, C., Boondaeng, A., Apiwatanapiwat, W., Kosugi, A., Arai, T., Sudesh, K., & Vaithanomsat, P. (2021). Enhanced Polyhydroxybutyrate (PHB) Production by Newly Isolated Rare Actinomycetes Rhodococcus sp. Strain BSRT1-1 Using Response Surface Methodology. *Scientific Reports, 11*(1), 1896. doi: 10.1038/s41598-021-81386-2

Verlinden, R. A. J. A. J., Hill, D. J. J., Kenward, M. A. A., Williams, C. D. D., & Radecka, I. (2007). Bacterial Synthesis of Biodegradable Polyhydroxyalkanoates. *Journal of Applied Microbiology, 102*(6), 1437–1449. doi: 10.1111/j.1365-2672.2007.03335.x

Witko, T. (2019). *Biophysical Characteristics and Cellular Studies of Polyhydroxyoctanoate (PHO) – Biodegradable and Biocompatible Polymer for Biomedical Applications.* Jagiellonian University, Krakow.

Zhu, Y., Ai, M., & Jia, X. (2022). Optimization of a Two-Species Microbial Consortium for Improved Mcl-PHA Production from Glucose–Xylose Mixtures. *Frontiers in Bioengineering and Biotechnology, 9*. doi: 10.3389/fbioe.2021.794331

14

COST-BENEFIT ANALYSIS IN THE BIOPOLYMER LIFECYCLE

Artur Jachimowski and Tomasz Witko

14.1   Introduction

Cost-benefit analysis is a comprehensive evaluation method that takes into account both quantitative and qualitative elements, as well as economic, social, and environmental aspects. It provides essential information on the benefits arising from a given situation and guidance on which variants to choose from among many solutions. Additionally, cost-benefit analysis serves as an aid when making informed decisions regarding the effectiveness of a specific project or determining whether an existing project is suitable. In most decision-making problems, two sides – benefits and costs – typically appear, which are often incompatible within a single hierarchical model. Therefore, this chapter presents the application of the modified analytical hierarchical process (AHP/ANP) method in solving multi-criteria decision-making (MCDM) problems in the context of supporting the design and modification of sustainable products, using biopolymers as an example.

It is worth emphasizing that cost-benefit analysis is extremely important in the context of designing and improving sustainable products, such as biopolymers, throughout their life cycle. It requires taking into account various aspects, ranging from economic production costs to social and ethical implications, and environmental impact. This analysis enables designers and decision-makers can make rational choices, taking into account not only short-term benefits but also the long-term consequences for society and the environment.

In the case of biopolymers, whose development is vital for the future of sustainable production, cost-benefit analysis makes it possible to assess the profitability of investing in research into new technologies or the development of alternative raw materials. Furthermore, it enables the identification of areas where further research and investment are needed to increase the competitiveness of biopolymers on the

DOI: 10.4324/9781032710693-15

market and minimize their environmental impact. Therefore, the application of the analytical network process (ANP) method in the case of biopolymers is not only justified but also essential for ensuring sustainable development in the chemical and packaging industries.

The proposed ANP models – benefits, costs, opportunities, and risks – include all kinds of dependencies and feedback between decision-making entities, reflecting the complexity of the decision-making problem and the relationships between factors inside the company and in its environment. The aim of this chapter is to develop models to help make the final decisions on the manufacture of sustainable products supported by logical arguments.

14.2   Decision-making in the biopolymer life cycle

Contemporary society is increasingly focusing its attention on issues of sustainable development and environmental protection. In response to growing concerns about plastic pollution, biodegradable and bio-based polymers, also known as biopolymers, are gaining in popularity (Możejko-Ciesielska & Kiewisz, 2016; Nitkiewicz et al., 2020). However, before this industry can truly contribute to improving the environment, a comprehensive cost-benefit analysis of biopolymers is necessary, as is shown in the literature (Ali et al., 2021; Fukuda & Kono, 2021; López-Marín et al., 2022). At this point, the authors will discuss the role of the decision-making process in evaluating the life cycle of biopolymers, elucidating both the significance of the decision-making process itself and the challenges associated with its application in the context of biopolymers.

The key step in the decision-making process is defining the objectives and scope of the analysis (Kim & Dale, 2005). In the case of biopolymers, these objectives may include identifying the main environmental impact areas, comparing biopolymers with traditional polymers, and assessing the potential benefits and costs associated with the use of biopolymers in various applications. Determining the scope of the analysis requires selecting a specific production system and identifying all stages of the biopolymer's life cycle, from raw material acquisition to recycling or disposal (Nitkiewicz et al., 2020).

14.2.1   The decision-making process during the biopolymer production stage

The decision-making process at the production stage of biopolymers involves a series of crucial steps aimed at ensuring efficient, sustainable, and environmentally friendly manufacturing practices. The first decision in the production of biopolymers involves choosing the raw materials. Biopolymers can be derived from various renewable sources, such as corn, sugarcane, potatoes, cellulose, or even algae. The decision-maker needs to consider factors such as availability, cost-effectiveness, environmental impact, and compatibility with the desired properties of the final biopolymer product.

There are several methods for producing biopolymers, including fermentation, enzymatic processes, and chemical synthesis. Each method has its advantages and disadvantages in terms of resource efficiency, energy consumption, waste generation, and environmental impact. The decision-maker must evaluate these factors to determine the most suitable production method for their specific needs (Sofińska et al., 2019; Witko et al., 2018).

Once the raw materials and production method are selected, the next step is to optimize the production process. This involves fine-tuning parameters such as temperature, pressure, pH, and reaction time to maximize efficiency, minimize waste, and reduce energy consumption. Process optimization may also include implementing recycling and reuse strategies to minimize environmental impact.

Ensuring the quality and consistency of biopolymer products is essential for their successful application. Decision-makers need to implement rigorous quality control measures at every stage of the production process to meet industry standards and customer requirements. This may involve testing for physical, chemical, mechanical, and biological properties to ensure that the biopolymers comply with performance specifications.

Throughout the production process, the environmental impact of biopolymer manufacturing must be continuously assessed. This step includes monitoring resource consumption, greenhouse gas emissions, water usage, waste generation, and other environmental indicators. By identifying areas of improvement and implementing sustainable practices, manufacturers can minimize their ecological footprint and enhance the overall sustainability of biopolymer production.

Compliance with relevant regulations and standards is essential for the production of biopolymers. Decision-makers must stay informed about local, national, and international regulations governing biopolymer manufacturing, including environmental regulations, safety standards, and labeling requirements. Failure to comply with these regulations can result in legal penalties, reputational damage, and market exclusion (Mtibe et al., 2021).

Finally, a cost-benefit analysis must be conducted to evaluate the economic viability of biopolymer production. This involves comparing the costs of raw materials, production processes, labor, and overheads with the potential benefits such as market demand, competitive advantage, and environmental benefits. By weighing up the costs and benefits, decision-makers can make informed choices about resource allocation, investment priorities, and business strategies (Patel & Blumberga, 2023).

14.2.2   The decision-making process during the biopolymer use phase

The decision-making process during the use phase of biopolymers constitutes a critical aspect of their overall life cycle assessment and sustainability. This phase encompasses the period when biopolymers are utilized in various applications, such as packaging, textiles, automotive parts, or biomedical devices. Effective decision-making during this stage involves taking into account factors related to product performance, environmental impact, and end-of-life options.

Assessing the performance of biopolymers in their intended applications is an essential step in this analysis. It involves considering such factors as mechanical properties, durability, barrier properties (e.g., moisture resistance, gas permeability), thermal stability, and compatibility with other materials (Cichoń et al., 2019; Solarz et al., 2023). Comparing the performance of biopolymers with that of conventional polymers or alternative materials is crucial for making informed decisions. Stakeholders need to ensure that the chosen biopolymer meets the required standards and specifications for the given application.

Monitoring and evaluating the environmental impact of biopolymer use throughout the lifecycle of this material is essential. This includes assessing factors such as energy consumption, greenhouse gas emissions, water usage, and potential pollution (Nessi et al., 2018).

Planning for the end-of-life management of biopolymers is a crucial aspect of decision-making during the use phase. Stakeholders need to consider options such as recycling, composting, or incineration. Evaluating the environmental impact and feasibility of different end-of-life options is essential. For example, composting biopolymers may be environmentally beneficial but requires appropriate infrastructure and conditions for effective decomposition (Kuruppalil, 2011).

An important step in the decision-making process is to identify and evaluate available end-of-life options for biopolymers. These may include composting, recycling, anaerobic digestion, incineration, or landfilling. Each option has its own environmental implications, cost considerations, and feasibility factors that need to be carefully assessed. Once the end-of-life options are identified, the next step is to conduct an environmental impact assessment for each option. This involves evaluating factors such as greenhouse gas emissions, energy consumption, water usage, and potential pollution associated with each disposal method. The goal is to determine which option minimizes environmental harm and maximizes resource efficiency. The decision-making process at the biopolymer disposal stage involves evaluating end-of-life options, conducting environmental and cost-benefit analyses, engaging stakeholders, ensuring regulatory compliance, and striving for continuous improvement. By carefully considering these factors, stakeholders can make informed decisions that maximize the sustainability of biopolymer disposal and contribute to a circular economy.

14.3 Analytic network process methodology

In this chapter, ANP was utilized to analyze the costs and benefits associated with the life cycle of biopolymers. ANP is an MCDM method proposed by Saaty. It is an advance on the earlier method called AHP. Both techniques are used to determine strategic priorities in various areas. Unlike AHP, which structures the decision

TABLE 14.1 The scale of comparisons in the ANP method

Degree of importance	Determination
1	Equal importance
3	The advantage of one element over the other is small
5	The advantage of one element over the other is moderate (significant)
7	The advantage of one element over the other is pronounced (very strong)
9	The advantage of one element over the other is very large (absolute)
2, 4, 6, 8	Intermediate values between odd ratings
Reverse values of the above numbers	If element *i* compared to element *j* receives one of the numbers mentioned above, then element *j* compared to element *i* receives the opposite value.

Source: Based on Saaty and Vargas (2001).

problem into a hierarchical structure comprising a goal, decision criteria, and alternatives, ANP treats it as a network (Saaty & Vargas, 2001). Both AHP and ANP utilize pairwise comparisons to measure the weights of structural elements and ultimately to rank alternatives in the decision-making process. ANP places a particular emphasis on the interaction and feedback between elements both within clusters and between clusters (Saputro et al., 2023). During the execution of ANP, several steps need to be taken, including model construction, creating pairwise comparison matrices between groups/clusters, and synthesizing priorities for each cluster. Model construction aims to define the problem and determine the criteria for the desired solution. The weights of components are determined using a quantitative scale from 1 to 9 (Table 14.1), which describes the level of importance of an element relative to others (Du & Sun, 2020; Saaty & Vargas, 2001).

The ANP process comprises the following key steps (Abdel-Baset et al., 2019; Saaty & Vargas, 2001; Saputro et al., 2023; Schulze-González et al., 2021; Utama et al., 2022):

1 **Problem definition** – initially, the decision problem is clearly outlined, and the main criteria, sub-criteria, and alternatives identified. This step also involves determining interactions between criteria, internal and external dependencies, and any feedback loops.

2 **Priority vector determination** – pairwise comparisons are conducted between criteria to establish priority vectors. Nodes within each cluster are compared based on their relationships, both internal and external, so as to gauge the distribution of influence across the network. These comparisons are made using expert-provided data.

3 **Consistency assessment** – the consistency of the comparison matrix obtained in the previous step is evaluated. Each matrix's consistency ratio (CR) is

calculated, with comparisons deemed consistent if the CR value is ≤0.10; otherwise, they require reviewing.

4 **Supermatrix formation** – a supermatrix is constructed, consisting of submatrices representing relationships between nodes within each cluster, as dictated by specific dependencies. Each part of the supermatrix illustrates the relationship between two factors within the system. The long-term relative impact of criteria on each other is determined by exponentiating the supermatrix.

5 **Global priority vector calculation** – the weighted supermatrix is transformed into a boundary supermatrix to establish a stable set of weights representing the relative influence of nodes on each other over the long term.

6 **Importance level determination** – the final stage involves assigning levels of importance (weights) to alternatives and criteria. The alternative with the highest weight in selection problems is considered the best, while in the case of weighting issues, the highest weight is deemed the most crucial criterion.

In the ANP, decision-making problems are depicted as a network consisting of criteria and alternatives, collectively referred to as elements, organized into groups. The relationships between these elements can vary, allowing for feedback loops and interdependencies within and between groups (Figure 14.1). This flexibility

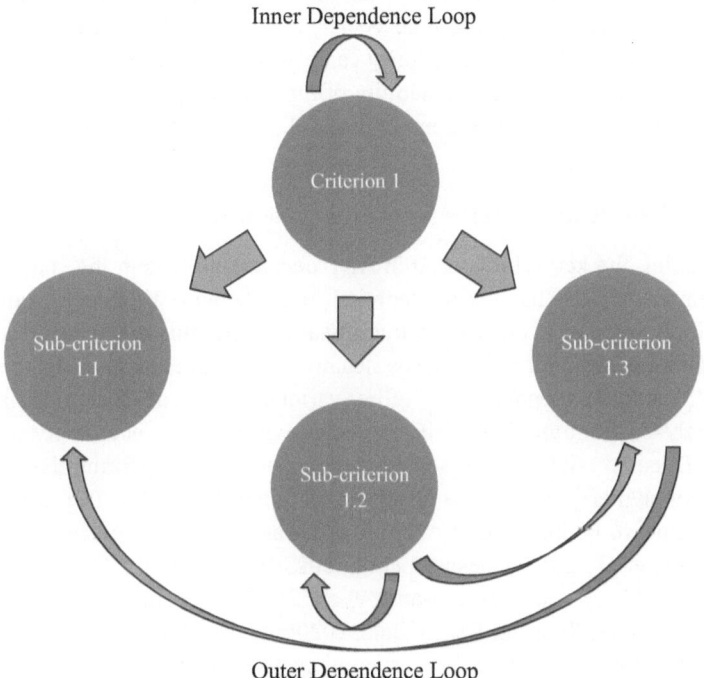

FIGURE 14.1 Interdependencies in ANP based on (Abdel-Baset et al., 2019).

allows the user to take into account interdependences between criteria, offering a more nuanced approach to modeling complex scenarios. The influence of elements on each other within the network can be illustrated using a supermatrix. This novel concept involves a two-dimensional matrix that adjusts the relative importance of weights from individual pairwise comparison matrices to create a new global supermatrix incorporating eigenvectors of the adjusted weights (Du & Sun, 2020; Fernandez Portillo et al., 2019; Saputro et al., 2023; Schulze-González et al., 2021).

The decision to implement a concept or project variant requires a thorough examination of various possibilities, taking into account both positive and negative aspects. The positive aspects include benefits and opportunities, while the negative aspects encompass costs and risks. Any evaluation of these elements can be challenging as they often cannot be clearly expressed in numerical terms. Therefore, to avoid errors, it is necessary to rely on proven mathematical methods, such as MCDM methods, which constitute a significant field of operations research. By means of such methods as ANP/AHP and BOCR analysis, it is possible to conduct critical validation of the decision problem. After making all pairwise comparisons, the results are synthesized within each control subsystem: benefits, opportunities, costs, and risks. Before selecting the best solution, the results obtained for the four control subsystems need to be combined. In the final phase of BOCR analysis, sensitivity analysis is also conducted for decision alternatives. This is a crucial element of the procedure that allows for the remaining variants to be taken into account when determining the final decision. Often, doubts arise over whether another solution with a similar priority can also be a good choice, and why. This can be determined based on the stability of the sensitivity function (Altan & Işık, 2023; Kovur & Gedela, 2020; Lammoglia et al., 2020; Petrillo et al., 2023).

14.4   Identification of criteria used in ANP models

At this point, the key criteria used in ANP decision models in the context of the biopolymer life cycle have been identified. Biopolymers, which are becoming an increasingly popular alternative to traditional polymer materials, attract attention due to their potential for reducing any negative environmental impact and promoting sustainable development. Therefore, various aspects of these materials have been analyzed from the perspective of their environmental, economic, and social impact. Detailed criteria for analyzing the benefits, costs, opportunities, and risks in the biopolymer life cycle are presented in Tables 14.2–14.5. The identification of the factors to be included in ANP models was based on a review of the literature (Chang et al., 2016; Chesneau et al., 2023; Dziuba et al., 2021; Gowthaman et al., 2021; Mtibe et al., 2021; Nanda et al., 2022; Ohkoshi et al., 2000; Patel & Blumberga, 2023; Rababah & AL- Oqla, 2020; Stevens, 2021; Tanase et al., 2014; Wellenreuther et al., 2022) and expert opinions from the field.

The above-presented criteria developed for ANP models are crucial for assessing the overall impact of biopolymers on the surrounding reality. In the environmental

TABLE 14.2 Criteria for benefits analysis in the biopolymer life cycle

Criteria	*Sub-criteria*	*Explanations*
Environmental	Reduction in greenhouse gas emissions	An evaluation of the ability of biopolymers to reduce greenhouse gas emissions compared to traditional petroleum-based polymers
	Biodegradation and recycling	An evaluation both of the ability of biopolymers to biodegrade and their recyclability to minimize negative environmental impact
	Reduction in natural resource consumption	An analysis of the reduction in natural resource consumption, such as crude oil and natural gas, through the use of biopolymers
	Biodiversity conservation	An assessment of the impact of biopolymer production and use on biodiversity conservation and ecosystem protection
	Alternative to plastic	An analysis of the effects of replacing traditional plastics with biopolymers to reduce the amount of plastic waste and mitigate issues associated with their accumulation in the environment
Economic	Price competitiveness	An analysis of the price competitiveness of biopolymers in the market compared with traditional polymers and their impact on consumer decisions
	Regulatory support	In some regions, governments may offer financial incentives, tax breaks, or other forms of support for the production and use of biopolymers, especially in the context of promoting more sustainable production practices
	New markets and segments	Biopolymers open up new markets and segments for companies, which can leverage the ecological and economic benefits of these materials to acquire new customers and boost their competitiveness
	Reducing dependence on petroleum resources	An analysis of the effects of reducing the risk associated with fluctuations in oil prices by using alternative resources for biopolymer production

(Continued)

TABLE 14.2 (Continued)

Criteria	Sub-criteria	Explanations
Social	Promotion of sustainable lifestyle	An assessment of the impact of biopolymer promotion on ecological awareness and encouraging a more sustainable lifestyle. Improving company image
	Improvements in company image	An evaluation of the benefits to the company's image resulting from the use of biopolymers as a more environmentally friendly and sustainable alternative
	Education and awareness	An assessment of educational activities and information campaigns aimed at increasing public awareness of the benefits of biopolymers
	Development of local communities	An analysis of the impact of biopolymer production on the development of local communities, their economy, and the standard of living of residents

TABLE 14.3 Cost analysis criteria in the biopolymer life cycle

Criteria	Sub-criteria	Explanations
Environmental	Costs associated with greenhouse gas emissions	Explaining the costs arising from emissions of CO_2 and other greenhouse gases during the production, transportation, use, and disposal of biopolymers
	Environmental degradation costs	Examining the costs associated with the degradation of the natural environment, such as water and soil pollution, caused by the production and disposal of biopolymers
	Costs of natural resource consumption	Including the costs associated with the consumption of natural resources during the production of biopolymers
Economic	Production costs	Assessing the costs associated with the biopolymer production process, including raw materials, energy, labor, and infrastructure investment
	Distribution and transportation costs	Analysing the costs associated with the transportation of biopolymers from the producer to the customer, as well as distribution costs
	Disposal and recycling costs	Evaluating the costs associated with the disposal, recycling, or composting of biopolymers after their use

(Continued)

TABLE 14.3 (Continued)

Criteria	Sub-criteria	Explanations
Social	Public health costs	Examining the costs associated with the impact of biopolymers on public health, such as treatment costs related to toxicity or allergenicity
	Social costs connected with employment	Assessing the costs associated with job creation, employee training, and personnel management in the biopolymer sector
	Adaptation and social acceptance costs	Assessing the costs of adapting the local community to changes resulting from the production and use of biopolymers, as well as the costs associated with building social acceptance for these technologies

TABLE 14.4 Opportunity analysis criteria in the biopolymer life cycle

Criteria	Sub-criteria	Explanations
Environmental	Sustainable management of natural resources	An opportunity to apply more sustainable practices in the utilization of natural resources, such as soil, water, and energy, during biopolymer production
	Improvement of air and water quality	An opportunity to improve air and water quality by using biopolymers that can reduce emissions of harmful substances or undergo biodegradation in natural conditions
	Renewable resources	An opportunity to utilize an increasing number of renewable sources of raw materials for biopolymer production, which may reduce pressure on the natural environment
Economic	Development of new markets	An opportunity to create new markets and the potential for increased demand for biopolymers in various industrial sectors
	Technological innovations	The possibility of developing innovative technologies for biopolymer production, which can reduce production costs and enhance their efficiency
	Diversification of raw material sources	An opportunity to diversify the sources of raw materials used in biopolymer production, which may increase flexibility and resilience to changes in raw material prices

(*Continued*)

TABLE 14.4 (Continued)

Criteria	Sub-criteria	Explanations
Social	Job creation	An opportunity to create new jobs connected with the production, processing, and recycling of biopolymers
	Increasing environmental awareness	A chance to raise public awareness of the benefits of using biodegradable biopolymers and their impact on the environment
	Improved relations with local communities	An opportunity to build better relationships with local communities by promoting sustainable development and environmental stewardship

TABLE 14.5 Risk analysis criteria in the life cycle of biopolymers

Criteria	Sub-criteria	Explanations
Environmental	Potential toxicity risk	An assessment of the risk associated with the release of toxic substances during the production, use, and disposal of biopolymers
	Environmental pollution risk	An assessment of the risk associated with potential contamination of water, soil, and air by substances used in the production, use, and disposal of biopolymers
	Ecosystem degradation risk	An examination of the risk of degradation and loss of biodiversity resulting from the production and use of biopolymers, including the risk of habitat loss and threats to fauna and flora
Economic	Investment risk	An evaluation of the risk associated with investments in biopolymer production technologies and their potential return on investments
	Variability of raw material prices	Identifying the risk associated with fluctuations in prices of biological raw materials on the global market
	Impact on competitiveness	An evaluation of the risk associated with the competitiveness of biopolymers on the market compared to petrochemical polymers

(Continued)

TABLE 14.5 (Continued)

Criteria	Sub-criteria	Explanations
Social	Health safety	An analysis of the risk associated with the potential impact of biopolymers on human health in production, use, and disposal processes
	Employment risk	Identifying the risk associated with potential changes in employment in the sector connected with biopolymer production
	Social acceptance	An assessment of the risk associated with the social acceptance of biopolymer use and potential societal reactions to their introduction on the market
	Availability for local communities	An analysis of the risk associated with the availability of biopolymers in local communities, in particular in developing countries

context, the analysis focuses on such aspects as reducing greenhouse gas emissions, natural resource consumption, and ecosystem degradation. Economic analysis, on the other hand, examines the costs of the production, processing, and disposal of biopolymers, as well as their impact on the profitability of companies and market development. Another important aspect is the social dimension of biopolymer use. The impact on consumers, their environmental awareness, and purchasing behavior is analyzed. Additionally, social analysis takes into account such factors as employment, the development of local communities, and the reputation of companies. Analyzing these criteria will ultimately lead to a better understanding of the comprehensive impact of biopolymers on the environment, economy, and society. By using ANP decision models, we will be able to provide an accurate assessment of various aspects of biopolymers and be able to make better informed decisions regarding their use.

14.5 Decision models in the biopolymer life cycle

This chapter presents decision-making models based on the analytic network process, which focus on analyzing the benefits, costs, opportunities, and risks associated with the life cycle of biopolymers, taking into account their impact on various environmental, economic, and social aspects (Figures 14.2–14.5). As a consequence, these three key criteria occupy a pivotal position in this analysis. In each model, environmental analysis focuses on the impact of biopolymers on the natural environment, including in terms of reducing greenhouse gas emissions,

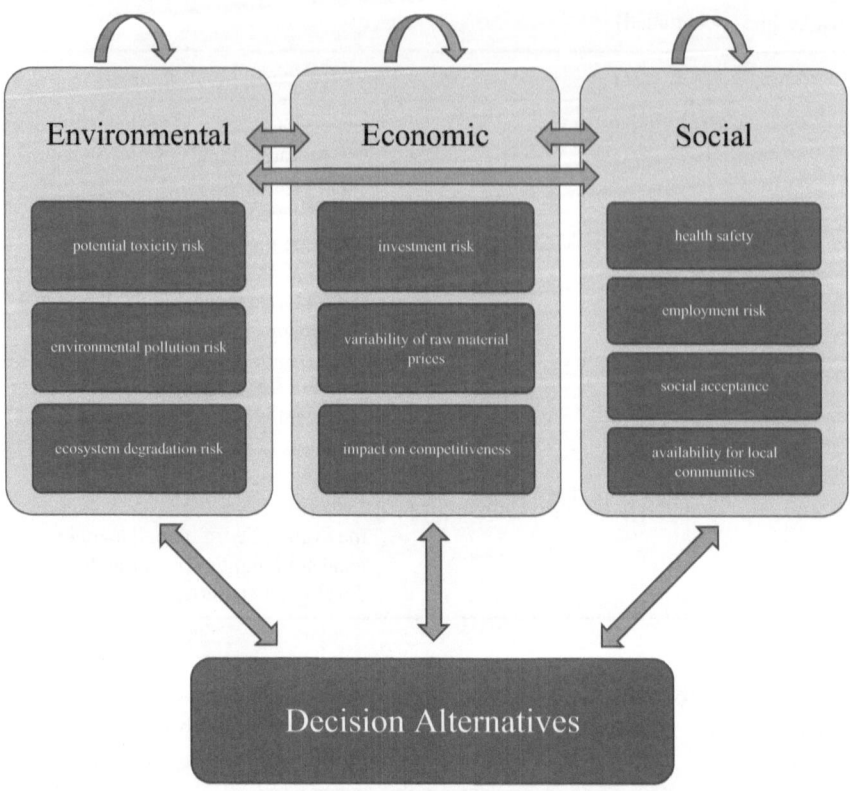

Environmental

potential toxicity risk

environmental pollution risk

ecosystem degradation risk

Economic

investment risk

variability of raw material prices

impact on competitiveness

Social

health safety

employment risk

social acceptance

availability for local communities

Decision Alternatives

FIGURE 14.2 ANP model of benefits in the biopolymer lifecycle.

natural resource consumption, and ecosystem degradation. Economic analysis focuses on the costs associated with the production, processing, and disposal of biopolymers, as well as on their impact on the profitability of businesses and market development. Meanwhile, social analysis evaluates the impact of biopolymers on consumers, their environmental awareness, purchasing behavior, as well as on employment, the development of local communities, and company reputation. By applying the proposed decision-making models based on ANP, we will be able to provide an accurate assessment of various aspects of biopolymers and support making more informed decisions regarding their use. These models will also provide a comprehensive analysis that takes into account sustainable development and long-term benefits for the environment, economy, and society.

Biopolymers offer several environmental advantages over traditional plastics. They are typically derived from renewable resources such as plants or algae, thus reducing reliance on finite fossil fuels. Additionally, biopolymers often biodegrade more easily than conventional plastics, potentially mitigating the accumulation of plastic waste in ecosystems and bodies of water (Witko, 2019). Furthermore, their production tends to generate fewer greenhouse gas emissions, contributing

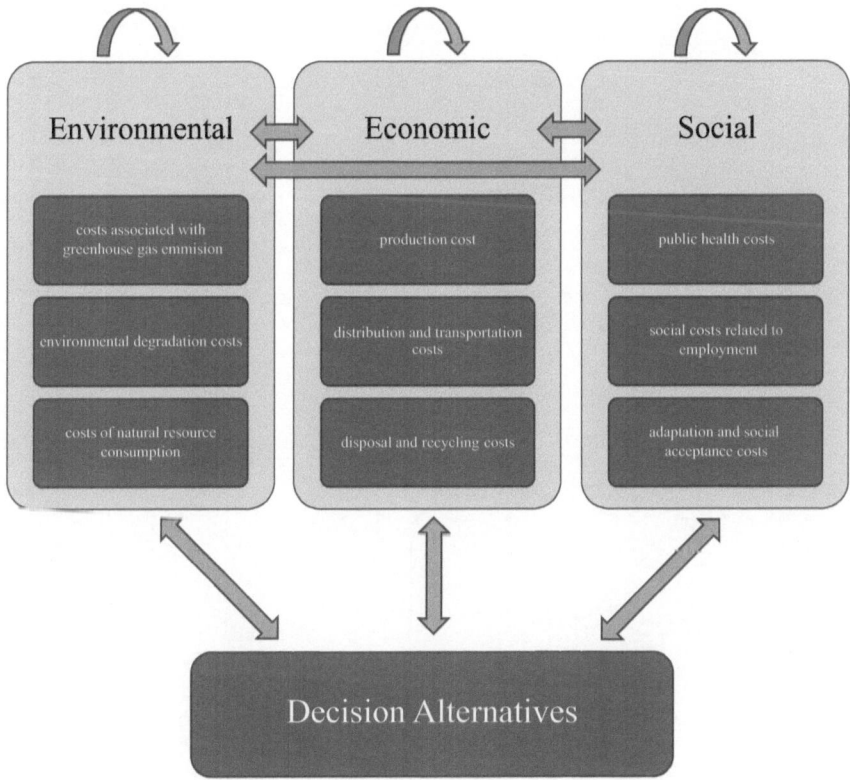

FIGURE 14.3 ANP model of costs in the biopolymer life cycle.

to overall carbon footprint reduction. The production of biopolymers can be more resource-efficient compared to traditional plastics. Many biopolymer feedstocks can be grown in agricultural settings, utilizing land that may not be suitable for food production. Additionally, advances in biopolymer manufacturing processes have led to improved efficiency in resource utilization, reducing energy consumption and waste generation during production. Biopolymers have the potential to facilitate a transition towards a circular economy model (Guzik et al., 2020). By means of composting or other biodegradation processes, biopolymers can be converted back into organic matter or energy, closing the loop on resource utilization. Additionally, biopolymers can be designed to be recyclable, enabling them to be incorporated into recycling streams and curtailing the need for virgin plastic production. Many governments and regulatory bodies are implementing policies encouraging the use of biodegradable and renewable materials (Dziuba et al., 2021). Businesses that adopt biopolymers may gain access to incentives, subsidies, or preferential treatment in procurement processes. Additionally, the use of biopolymers can help companies align with environmental regulations and meet sustainability goals, thereby reducing the risk of non-compliance penalties. The

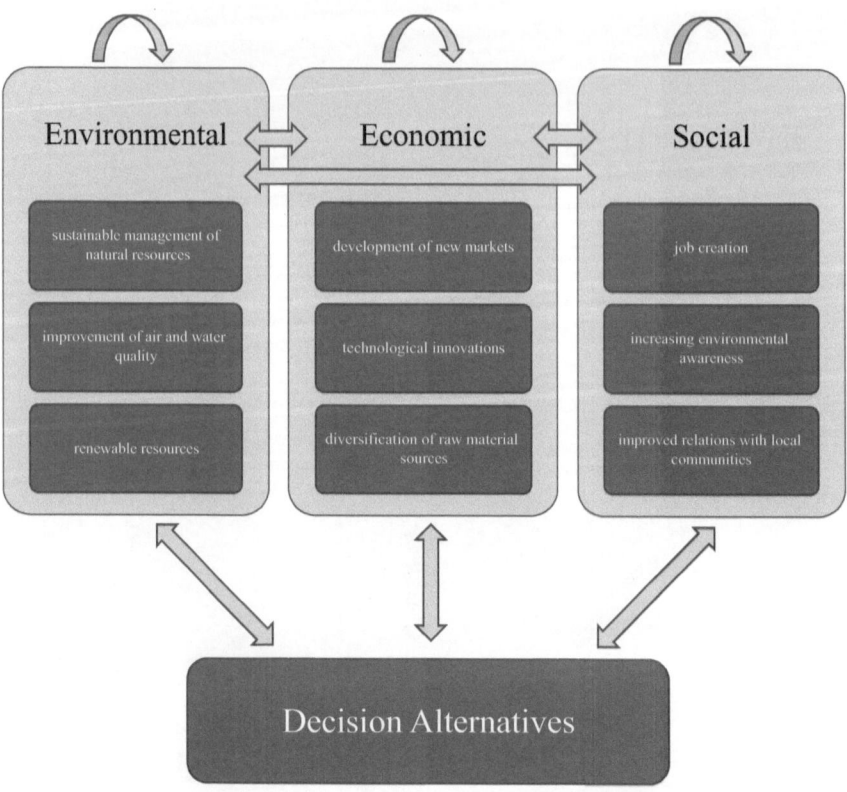

FIGURE 14.4 ANP model of opportunities in the biopolymer life cycle.

benefit analysis of biopolymers demonstrates their potential to make a major positive impact on environmental protection, resource efficiency, waste reduction, market opportunities, and regulatory compliance. By harnessing these benefits, the adoption of biopolymers as a sustainable alternative to traditional plastics can be promoted and accelerated.

In the cost analysis phase, several key factors need to be evaluated to assess the financial implications of biopolymer production and management. Production costs include all the expenses associated with acquiring raw materials, conducting manufacturing processes, labor costs, energy consumption, and overhead expenses required for biopolymer production. Evaluating the market dynamics is crucial for gauging demand for biopolymers, pricing trends, competition compared to traditional plastics, and potential revenue streams derived from selling biopolymers or their by-products. An analysis of the expenses connected with the disposal of biopolymers is also essential. They include the costs incurred in collecting, transporting, and processing biopolymers by means of various disposal methods such as composting, recycling, or incineration. Each of these disposal options carries its own set of costs, and analyzing them helps organizations make informed decisions

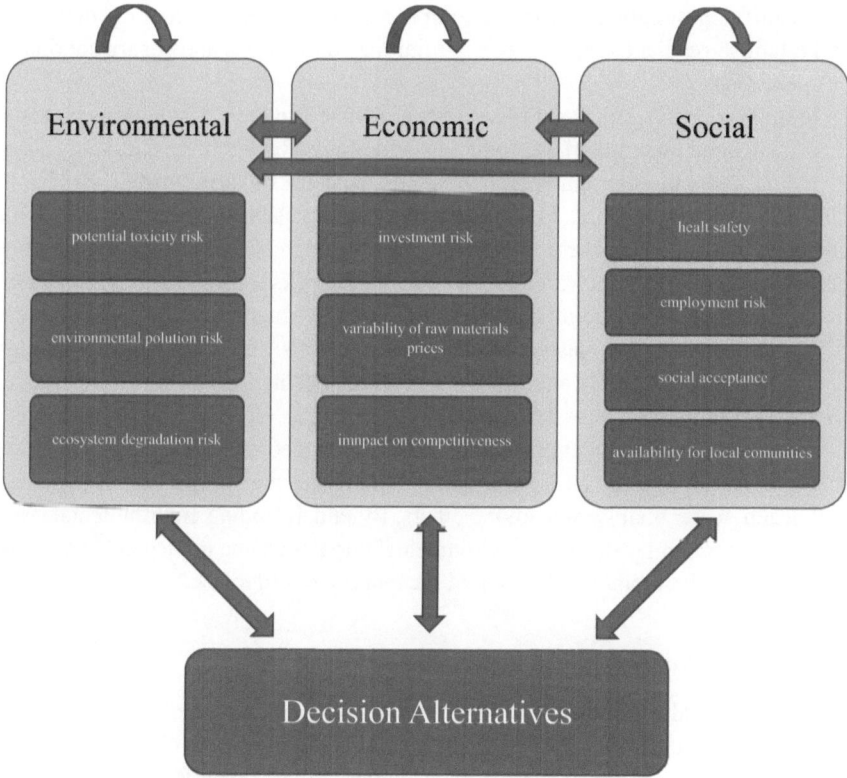

FIGURE 14.5   ANP risk model in the biopolymer life cycle.

regarding the most economically feasible end-of-life management strategy for biopolymers.

In the chance analysis phase, numerous factors need to be examined to understand the potential opportunities and risks associated with biopolymers. This includes an analysis of market trends, consumer preferences, and regulatory developments affecting biopolymers. It also entails monitoring changes in demand, advances in technology, and government policies supporting sustainable materials. Understanding these trends helps organizations identify market opportunities and adapt strategies accordingly. Assessing the likelihood of technological advances is key to gauging the future prospects of biopolymers. This analysis focuses on identifying innovations that could enhance production efficiency, improve performance attributes, and increase the cost-effectiveness of biopolymers over time. Anticipating technological breakthroughs enables businesses to stay competitive and capitalize on emerging opportunities. Assessing risks connected with the supply chain is essential to mitigate potential disruptions and uncertainties. This stage involves evaluating the availability and price volatility of raw materials, potential disruptions in production processes, and challenges in logistics and distribution.

By identifying supply chain risks, companies can implement contingency plans and establish resilient supply chains to minimize disruptions and ensure continuity of operations.

In the risk analysis phase, it is essential to identify and evaluate various potential risks associated with materials production and usage. Recognizing regulatory risks, including shifts in environmental regulations, changes in waste management policies, and evolving standards for compostability or recyclability is a crucial part of the analysis. Understanding these risks helps ensure regulatory compliance and affective adjustment to regulatory changes. Assessing market risks is also important for anticipating and mitigating potential challenges caused by fluctuations in demand, competition from traditional plastics or alternative materials, and uncertainties in pricing and market acceptance. By analyzing market risks, businesses can develop strategies to address competitive pressures and market dynamics effectively. Assessing environmental risks entails identifying potential unintended consequences of biopolymer production or disposal. This includes identifying their impact on ecosystems, soil health, water quality, and among others. By understanding environmental risks, measures to minimize negative environmental impacts can be implemented and sustainability can be promoted throughout the biopolymer lifecycle.

14.6   Conclusions

Cost-benefit analysis in the biopolymer lifecycle plays a crucial role when assessing the sustainable utilization of these materials (Pathak et al., 2014). By developing four ANP models – benefits, costs, opportunities, and risks – it is possible to gain a comprehensive understanding of various aspects of sustainable products. The benefits model makes it possible to identify and evaluate the positive effects of biopolymer application, such as lower CO_2 emissions, reduced fossil fuel consumption, and enhanced recycling and biodegradation, thereby contributing to sustainable development. Meanwhile, cost analysis takes into account both direct and indirect costs associated with the production, distribution, use, and disposal of biopolymers. Additionally, it considers social costs, such as the impact on health and the environment, thus enabling a comprehensive economic assessment. The opportunities model focuses on the potential for development and innovation in biopolymers, helps identify potential benefits, such as the development of new production technologies, identifies new application markets, and improves recycling and biodegradation processes. Conversely, the risk model allows for the identification and management of various types of risk, such as market uncertainty, raw material price volatility, and risks to health and the environment associated with new materials. In summary, the integration of these four ANP models provides a coherent analysis of the benefits, costs, opportunities, and risks in the lifecycle of biopolymers. This, in turn, enables more informed decisions regarding their production, utilization, and waste management, taking into account various environmental, economic, and social aspects. Therefore, it is imperative that further improvements be made in cost-benefit analysis in the biopolymer lifecycle, encompassing various aspects of sustainable utilization of these materials, such as their impact on the environment, health, and economy.

Acknowledgement

The publication/article presents the result of the Project no 070/ZJE/2024/POT financed from the subsidy granted to the Krakow University of Economics.

References

Abdel-Baset, M., Chang, V., Gamal, A., & Smarandache, F. (2019). RETRACTED: An integrated neutrosophic ANP and VIKOR method for achieving sustainable supplier selection: A case study in importing field. *Computers in Industry, 106*, 94–110. doi: 10.1016/j.compind.2018.12.017

Ali, Y., Sara, S., & Rehman, O. ur. (2021). How to tackle plastic bags and bottles pollution crisis in Pakistan? A cost–benefit analysis approach. *Environmental and Ecological Statistics, 28*(3), 697–727. doi: 10.1007/s10651-021-00511-6

Altan, E., & Işık, Z. (2023). Digital twins in lean construction: A neutrosophic AHP – BOCR analysis approach. *Engineering, Construction and Architectural Management*. doi: 10.1108/ECAM-11-2022-1115

Chang, I., Im, J., & Cho, G.-C. (2016). Introduction of microbial biopolymers in soil treatment for future environmentally-friendly and sustainable geotechnical engineering. *Sustainability, 8*(3), 251. doi: 10.3390/su8030251

Chesneau, C., Larue, L., & Belbekhouche, S. (2023). Design of tailor-made biopolymer-based capsules for biological application by combining porous particles and polysaccharide assembly. *Pharmaceutics, 15*(6), 1718. doi: 10.3390/pharmaceutics15061718

Cichoń, E., Haraźna, K., Skibiński, S., Witko, T., Zima, A., Ślósarczyk, A., Zimowska, M., Witko, M., Leszczyński, B., Wróbel, A., & Guzik, M. (2019). Novel bioresorbable tricalcium phosphate/polyhydroxyoctanoate (TCP/PHO) composites as scaffolds for bone tissue engineering applications. *Journal of the Mechanical Behavior of Biomedical Materials, 98*, 235–245. doi: 10.1016/j.jmbbm.2019.06.028

Du, Y.-W., & Sun, Y.-L. (2020). DS/ANP method: A simplified group analytic network process with consensus reaching. *IEEE Access, 8*, 35726–35741. doi: 10.1109/ACCESS.2020.2972924

Dziuba, R., Kucharska, M., Madej-Kiełbik, L., Sulak, K., & Wiśniewska-Wrona, M. (2021). Biopolymers and biomaterials for special applications within the context of the circular economy. *Materials, 14*(24), 7704. doi: 10.3390/ma14247704

Fernandez Portillo, L. A., Nekhay, O., & Estepa Mohedano, L. (2019). Use of the ANP methodology to prioritize rural development strategies under the LEADER approach in protected areas. The case of Lagodekhi, Georgia. *Land Use Policy, 88*, 104121. doi: 10.1016/j.landusepol.2019.104121

Fukuda, K., & Kono, H. (2021). *Cost-Benefit Analysis and Industrial Potential of Exopolysaccharides* (pp. 303–339). doi: 10.1007/978-3-030-75289-7_12

Gowthaman, N. S. K., Lim, H. N., Sreeraj, T. R., Amalraj, A., & Gopi, S. (2021). Advantages of biopolymers over synthetic polymers. In *Biopolymers and Their Industrial Applications* (pp. 351–372). Elsevier. doi: 10.1016/B978-0-12-819240-5.00015-8

Guzik, M., Witko, T., Steinbüchel, A., Wojnarowska, M., Sołtysik, M., & Wawak, S. (2020). What has been trending in the research of polyhydroxyalkanoates? A systematic review. *Frontiers in Bioengineering and Biotechnology, 8.* doi: 10.3389/fbioe.2020.00959

Kim, S., & Dale, B. E. (2005). Life cycle assessment study of biopolymers (polyhydroxy-alkanoates) derived from no-tilled corn. *International Journal of Life Cycle Assessment, 10*(3), 200–210. doi: 10.1065/lca2004.08.171

Kovur, K. M., & Gedela, R. K. (2020). *An Integrated Approach of BOCR Modeling Framework for Decision Tool Evaluation* (pp. 109–148). doi: 10.1007/978-3-030-36518-9_5

Kuruppalil, Z. (2011). Green plastics: An emerging alternative for petroleum-based plastics. *International Journal of Engineering Research & Innovation, 3*(1), 59–64. doi: 978-1-60643-379-9

Lammoglia, J. A. D. M., Brandalise, N., & Hernandez, C. T. (2020). Analytical hierarchy process – Bocr applied for the best lean project selection for production lines. *Independent Journal of Management & Production, 11*(1), 054. doi: 10.14807/ijmp.v11i1.990

López-Marín, J., Gálvez, A., del Amor, F. M., Piñero, M. C., & Brotons-Martínez, J. M. (2022). The cost-benefits and risks of using Raffia made of biodegradable polymers: The case of pepper and tomato production in greenhouses. *Horticulturae, 8*(2), 133. doi: 10.3390/horticulturae8020133

Możejko-Ciesielska, J., & Kiewisz, R. (2016). Bacterial polyhydroxyalkanoates: Still fabulous? *Microbiological Research, 192*, 271–282. doi: 10.1016/j.micres.2016.07.010

Mtibe, A., Motloung, M. P., Bandyopadhyay, J., & Ray, S. S. (2021). Synthetic biopolymers and their composites: Advantages and limitations—An overview. *Macromolecular Rapid Communications, 42*(15). doi: 10.1002/marc.202100130

Nanda, S., Patra, B. R., Patel, R., Bakos, J., & Dalai, A. K. (2022). Innovations in applications and prospects of bioplastics and biopolymers: A review. *Environmental Chemistry Letters, 20*(1), 379–395. doi: 10.1007/s10311-021-01334-4

Nessi, S., Bulgheroni, C., Konti, A., Sinkko, T., Tonini, D., & Pant, R. (2018). Environmental sustainability assessment comparing through the means of lifecycle assessment the potential environmental impacts of the use of alternative feedstock (biomass, recycled plastics, CO2) for plastic articles in comparison to using [...] – part1. *JRC Technical Reports. European Commission, 268.* doi: 10.2760/XXXXX

Nitkiewicz, T., Wojnarowska, M., Sołtysik, M., Kaczmarski, A., Witko, T., Ingrao, C., & Guzik, M. (2020). How sustainable are biopolymers? Findings from a life cycle assessment of polyhydroxyalkanoate production from rapeseed-oil derivatives. *Science of the Total Environment, 749*, 141279. doi: 10.1016/j.scitotenv.2020.141279

Ohkoshi, I., Abe, H., & Doi, Y. (2000). Miscibility and solid-state structures for blends of poly[(S)-lactide] with atactic poly[(R,S)-3-hydroxybutyrate]. *Polymer, 41*(15), 5985–5992. doi: 10.1016/S0032-3861(99)00781-8

Patel, N., & Blumberga, D. (2023). Insights of bioeconomy: Biopolymer evaluation based on sustainability criteria. *Environmental and Climate Technologies, 27*(1), 323–338. doi: 10.2478/rtuect-2023-0025

Pathak, S., Sneha, C., & Baby Mathew, B. (2014). Bioplastics: Its timeline based scenario & challenges. *Journal of Polymer and Biopolymer Physics Chemistry, 2*(4), 84–90. doi: 10.12691/jpbpc-2-4-5

Petrillo, A., Salomon, V., & Tramarico, C. (2023). State-of-the-art review on the analytic hierarchy process with benefits, opportunities, costs, and risks. *Journal of Risk and Financial Management, 16*(8), 372. doi: 10.3390/jrfm16080372

Rababah, M. M., & AL- Oqla, F. M. (2020). Biopolymer composites and sustainability. In *Advanced Processing, Properties, and Applications of Starch and Other Bio-Based Polymers* (pp. 1–10). Elsevier. doi: 10.1016/B978-0-12-819661-8.00001-9

Saaty, T. L., & Vargas, L. G. (2001). *Models, Methods, Concepts & Applications of the Analytic Hierarchy Process* (Vol. 34). Boston, MA: Springer US. doi: 10.1007/978-1-4615-1665-1

Saputro, K. E. A., Hasim, Karlinasari, L., & Beik, I. S. (2023). Evaluation of sustainable rural tourism development with an integrated approach using MDS and ANP methods: Case study in ciamis, West Java, Indonesia. *Sustainability*, *15*(3), 1835. doi: 10.3390/su15031835

Schulze-González, E., Pastor-Ferrando, J.-P., & Aragonés-Beltrán, P. (2021). Testing a recent DEMATEL-based proposal to simplify the use of ANP. *Mathematics*, *9*(14), 1605. doi: 10.3390/math9141605

Sofińska, K., Barbasz, J., Witko, T., Dryzek, J., Haraźna, K., Witko, M., Kryściak-Czerwenka, J., & Guzik, M. (2019). Structural, topographical, and mechanical characteristics of purified polyhydroxyoctanoate polymer. *Journal of Applied Polymer Science*, *136*(4), 47192. doi: 10.1002/app.47192

Solarz, D., Witko, T., Karcz, R., Malagurski, I., Ponjavic, M., Levic, S., Nesic, A., Guzik, M., Savic, S., & Nikodinovic-Runic, J. (2023). Biological and physiochemical studies of electrospun polylactid/polyhydroxyoctanoate PLA/P(3HO) scaffolds for tissue engineering applications. *RSC Advances*, *13*(34), 24112–24128. doi: 10.1039/D3RA03021K

Stevens, C. V. (2021). *Bio-Based Packaging* (S. M. Sapuan & R. A. Ilyas (eds.)). Wiley. doi: 10.1002/9781119381228

Tanase, M., Rapa, M., & Popa, O. (2014). Biopolymers based on renewable resources – A review. *Scientific Bulletin. Series F. Biotechnologies*, *XVIII*(1), 11–20. doi: 10.2174/1874123101004010011

Utama, D. M., Parameswari, R. P., & Mubin, A. (2022). Evaluation and performance analysis using ANP and TOPSIS algorithm. *Journal of Physics: Conference Series*, *2394*(1), 012005. doi: 10.1088/1742-6596/2394/1/012005

Wellenreuther, C., Wolf, A., & Zander, N. (2022). Cost competitiveness of sustainable bioplastic feedstocks – A Monte Carlo analysis for polylactic acid. *Cleaner Engineering and Technology*, *6*, 100411. doi: 10.1016/j.clet.2022.100411

Witko, T. (2019). *Biophysical Characteristics and Cellular Studies of Polyhydroxyoctanoate (PHO) – Biodegradable and Biocompatible Polymer for Biomedical Applications*. Krakow: Jagiellonian University.

Witko, T., Guzik, M., Sofińska, K., Stepien, K., & Podobinska, K. (2018). Novel biocompatible polymers for biomedical applications. *Biophysical Journal*. doi: 10.1016/j.bpj.2017.11.2014

INDEX

Note: **Bold** page numbers refer to tables; *italic* page numbers refer to figures and page numbers followed by "n" denote endnotes.